▶ Christian WITTRISCH
▶ Henri CHOLET
IFP Energies nouvelles

PROGRESSING CAVITY PUMPS

Oil Well Production Artificial Lift

Second Edition
Revised and Expanded

2013

Editions TECHNIP 5 av. de la République, 75011 PARIS, FRANCE

FROM THE SAME PUBLISHER

- Heavy Crude Oils
  From Geology to Upgrading. An Overview
  A.Y. HUC

- Multiphase Production
  Pipeline Transport, Pumping and Metering
  J. FALCIMAIGNE, S. DECARRE

- Corrosion and Degradation of Metallic Materials
  Understanding of the Phenomena and Applications in Petroleum and Process Industries
  F. ROPITAL

- A Geoscientist's Guide to Petrophysics
  B. ZINSZNER, F.M. PERRIN

- Physico-Chemical Analysis of Industrial Catalysts
  A Practical Guide to Characterisation
  J. LYNCH

- Petroleum Microbiology (2 vols.)
  J.P. VANDECASTEELE

- Acido-Basic Catalysis (2 vols.)
  Application to Refining and Petrochemistry
  C. MARCILLY

- CO_2 Capture
  Technologies to Reduce Greenhouse Gas Emissions
  J. LECOMTE, P. BROUTIN, E. LEBAS

- Chemical Reactors
  From Design to Operation
  P. TRAMBOUZE, J.P. EUZEN

- The Geopolitics of Energy
  J.P. FAVENNEC

- Marine Oil Spills and Soils Contaminated by Hydrocarbons
  Environmental Stakes and Treatment of Pollutions
  C. BOCARD

- Well Production Practical Handbook
  H. CHOLET

All rights reserved.

No part of this publication may be reproduced or transmitted in any form or by any means, electronic or mechanical, including photocopy, recording, or any information storage and retrieval system, without the prior written permission of the publisher.

© Editions Technip, Paris, 2013.

Printed in France

ISBN 978-2-7108-0998-2

Preface

René Moineau was not only a visionary inventor when he filed the patent for a new gear mechanism in 1930 that could be used as a pump, motor or transmission device. He also set up a company – "Pompe Compresseur Mécanique" (PCM) – to manufacture and market his invention. Progressing cavity pumps are now widely used by the industry to transfer many types of fluids, as well as in artificial oil lifts.

A long story with a happy ending. During the 1980s, the Institut Français du Pétrole (now IFP Energies nouvelles) based in Rueil Malmaison (France) in partnership with PCM and associated with french petroleum companies took over leadership of the research, development and field testing of progressing cavity pumps applied to downhole artificial lift oil wells. IFP Energies nouvelles successfully ran the first "Moineau pump". Now known as "Progressing Cavity Pumps" or PCPs, they are installed in the bottom of the well and driven from the surface by solid rods in order to pump fluids. This opened the way for a new artificial lift method and equipment using a simple, economical and rugged down-hole pump. Many field tests have been carried out, proving that PCPs are suitable for many environments and different types of fluids, and can be used with heavy viscous oils containing gas and sand or light and abrasive oils. They are also suitable for use with today's high-temperature fluids from thermal recovery and high flow rates.

This new edition has come about thanks to Henry Cholet and Christian Wittrisch. It serves as an update to Henri Cholet's first book – "Progressing Cavity Pumps" – published by *Editions Technip* in 1997 and is designed to serve the same purpose: to provide concise and useful information on the main principles, features, performance and operational conditions. Within this framework, this second edition includes information about new features related to operating conditions, such as high flow rates, multiphase fluids, high temperature conditions, heavy oil processes, water flooding, dewatering gas well, monitoring of well production, etc.

Henri Cholet, who is now retired, was a project manager with a particularly high level of technical proficiency and a great deal of experience in improving pump technologies.

Christian Wittrisch is an IFP Energies nouvelles research engineer with wide-ranging experience in down-hole equipment – experience put for the purposes of updating the first edition of this book.

As Head of the Mechanical Engineering Department within the Applied Mechanics Division at IFP Energies nouvelles, I would like to thank Henri Cholet and Christian Wittrisch for their work in bringing about a better understanding of Progressing Cavity Pumps and for their help in marketing development, which enhances IFP Energies nouvelles' reputation.

Pascal Longuemare
Head of the Mechanical Engineering Department
within the Applied Mechanics Division at IFP Energies nouvelles

Foreword

Human genius has a tendency to look into apparently unsolvable problems and strive to develop solutions to them.

This was particularly the case with René Moineau, a famous First World War pilot and inventor of many advanced technologies, most of which were for use with aeroplanes.

René Moineau was a visionary inventor with a superb intellect and an inventive mind who would adopt a basic mechanical approach to problem-solving.

He built pump components with sheets of paper in order to manage his mathematical and geometric intuition and in 1930 filed the patent for a new gear mechanism that could be used as a pump, motor or transmission device.

In 1932 René Moineau, helped by the Gevelot Company, became the founding chairman of a new company: "Pompe Compresseur Mécanique", known today across the world as PCM.

The Progressing Cavity Pump was a success story from the start. Integrating a new concept and patented invention, it was rapidly adopted throughout the industry for transferring many types of fluid. During the 1980s, progressing cavity pumps started to be used to lift oil wells and are used worldwide today.

René Moineau's PCP invention had been applied and accepted by the market. It became a great innovation for the whole industry.

The objective of this new edition is to add many improvements to and uses for the PCP in oil wells. It can now be used in difficult environments, with different types of fluid, in high temperatures, and in conjunction with recently developed technologies for advanced materials, on driving heads, monitoring and safety control with data transmission. These new technologies for materials and data transmission and processing mean that the PCP operating life can be increased, and well production involves lower operating costs.

Our hope is that this new edition of the Progressing Cavity Pumps book for oil well artificial lift will help ensure that the Progressing Cavity Pump becomes better known, better use, more efficiency and ensure it a trustingly development.

Christian Wittrisch & Henri Cholet

Table of Contents

Preface V

Foreword VII

Chapter 1
WHAT IS A PROGRESSING CAVITY PUMP? WHAT IS IT USED FOR? WHAT NEW FEATURES DOES IT HAVE?

1.1	Introduction to the Progressing Cavity Pump	2
1.2	Moineau: the Pump Success Story for the Oil Industry	5
1.3	PCP Basic Principle	6

Chapter 2
PROGRESSING CAVITY PUMP: OPERATING PRINCIPLE

2.1	The PCP Principle.		9
2.2	PCP Geometry.		10
2.3	Rotor-Stator Geometries		12
	2.3.1	Theoretical Elements.	12
2.4	Mono and Multi-Lobes Pump Geometries		14
	2.4.1	Mono-Lobe (1-2 Pump) Geometry	15
	2.4.2	Multi-Lobes Pump Geometry.	19
	2.4.3	Multi-Lobe Geometry Pump's Applications	21
2.5	Pump Manufacturing Process		21
	2.5.1	Rotor Manufacturing Process.	21
	2.5.2	Stator Manufacturing Process.	22
2.6	Pump Rotation Transmission.		22
2.7	PCP and Oil Well Applications		22

Chapter 3
PCP CHARACTERISTICS

3.1	**Mono-Lobe Pump Characteristics**		25
	3.1.1	Pump Kinematics	25
	3.1.2	Pump Volume Displacement	26
	3.1.3	Pump Head Rating	26
	3.1.4	Pump Torque	27
	3.1.5	Load on Thrust Bearing Drive Head	28
3.2	**Multi-Lobe Pump Characteristics**		28
	3.2.1	Theoretical Rotating Speed	28
	3.2.2	Flow Rate	29
	3.2.3	Pump Torque	29
3.3	**Elastomers Stator Characteristics**		29
	3.3.1	Elastomer Selection *versus* Pumped Fluids	29
	3.3.2	Measurements Characterising an Elastomer	30
	3.3.3	Elastomer Characterization and Classification	31
	3.3.4	Elastomer Resistance Challenge	32
	3.3.5	Elastomer Conclusions	33
3.4	**Pump Efficiency Evaluation**		33
	3.4.1	Calculation of Volumetric Pump Efficiency	33
	3.4.2	Evaluation of the Pump's Actual Flow Rate	33
	3.4.3	Internal Recirculation Flow or Slippage Under Pressure	33
	3.4.4	Pump Efficiency and Fluid Slippage	34

Chapter 4
OIL CHARACTERISTICS

4.1	**API Oil Gravity Scale**	37
4.2	**Oil Viscosity and Temperature**	39
4.3	**Heavy Oil and Dilution**	40
4.4	**Heavy Oil and PCP**	41

Chapter 5
ARTIFICIAL LIFT AND SELECTION OF A PCP

5.1	The Reservoir Characteristics		43
5.2	The Produced Effluent Characteristics		44
5.3	Well Geometry, Architecture Drainage, Horizontal Well		45
5.4	Well Completion and Sand Control Liner Screens		45
5.5	PCP Sized to the Well Geometry and Completion		46
5.6	PCP's Operating Conditions		46
	5.6.1	Frictional Pressure Drop Generated by Viscosity	46
	5.6.2	Resistant Torque Generated by the Viscosity	47

5.7	**Fluid Viscosity Reducer**		48
5.8	**Fluid Formation and Gas Content**		48
5.9	**Fluid Temperature**		50
5.10	**Fluid and Presence of Sand**		50
	5.10.1	Sand Transportation in Production	50
	5.10.2	Incidence on the Pump	50
	5.10.3	Application Limits	51
5.11	**Production Criteria**		52
	5.11.1	Positioning Level According to the Dynamic Level (or Submergence Level)	52
	5.11.2	Positioning Level in Respect of the Bubble Point Level	52
	5.11.3	Evaluation of the Pump's Minimum Head Rating	53
	5.11.4	Pressure Generated by the Column Height to be Discharged	54
	5.11.5	Pressure Drop Generated by the Viscosity of the Effluent	54
	5.11.6	Wellhead Pressure	54
	5.11.7	Head of Pressure Related to Down Hole Artificial Lift	55
	5.11.8	Head of Pressure of Water and Fluid formulas	56

Chapter 6
OPERATIONAL CONDITIONS

6.1	**Flow Rate and Head Rating**		57
6.2	**Rotor Rotating Speed**		58
6.3	**Performance Controls and Tests**		59
	6.3.1	Test Procedure	59
	6.3.2	Test Report	59
6.4	**PCP Guideline and Identification**		59
	6.4.1	Pump Terms and Selection	60
	6.4.2	PCP Rotor and Stator Identifications	60
	6.4.3	PCP Manufacturers' Model Identification	61
6.5	**PCP Data Sheet from Manufacturers**		63
6.6	**Well Data Sheet**		66
	6.6.1	Well Situation	66
	6.6.2	Well Completion Data	66
	6.6.3	Well Production Data	66
	6.6.4	Production Fluid Data	68
	6.6.5	Well and Fluids Characteristics Influence Pump Performance	69
	6.6.6	Operational Specifications	69
	6.6.7	Physicochemical Treatment Programme	70
	6.6.8	PCP Data Sheet	70

Chapter 7
PRESENCE OF GAS AT THE PUMP INLET

7.1	**Vertical or Slightly Deviated Well**	71
7.2	**Highly Deviated or Horizontal Well**	71
	7.2.1 Permanent and Homogeneous Flow	71
	7.2.2 Stratified or Intermittent Flow	73
7.3	**A "Natural" Gas Separator**	73
7.4	**Static Gas Separator**	73
	7.4.1 Static Continuous Downhole Centrifugal Gas Separator	73
	7.4.2 System Avoiding Gas Suction in the Horizontal Wells	75
7.5	**Gor Calculation at the Pump Inlet**	77
	7.5.1 Solution Gas/Oil Ratio	77
	7.5.2 Gas Volume Factor	78
	7.5.3 Formation Volume Factor	78
	7.5.4 Total Volume of Fluids	79
7.6	**An Application Example**	80

Chapter 8
DRIVE RODS

8.1	**Solid Drive Rod Characteristics**	81
	8.1.1 Influence on the Tubing Diameter	83
	8.1.2 Drive Rod Characteristics	83
	8.1.3 Maximum Service Rod Torques	84
8.2	**Stresses in the Drive String**	84
	8.2.1 Weight of Solid Rods and Hollow Rods	84
	8.2.2 Thrust Generated by the Head Rating of the Pump	85
	8.2.3 Mechanical Resistant Torque	85
	8.2.4 Resistant Torque by the Effluent Viscosity in the Tubing	85
	8.2.5 Total Stresses in the Drive Strings	86
8.3	**Tubings and Drive String**	86
	8.3.1 Tubing and Centralizers	86
	8.3.2 Drive Rod String and NON Rotating Centralizers	87
	8.3.3 Case of Deviated and Horizontal Wells	88
	8.3.4 Contact Between the Tubing and the Drive Strings	88
8.4	**Hollow Rod Drive PCP**	91
	8.4.1 Tubing Drive PCP's	91
	8.4.2 Hollow Drive Rod	91
8.5	**Coiled Tubing Drive PCP**	93
8.6	**Continuous Drive Rod**	94
8.7	**Polished Rod and Clamp**	95
8.8	**Tubing Anchor**	95
	8.8.1 Tubing Anchor Catcher	95
	8.8.2 Tubing Torque Anchor	96

8.9	**Other Downhole Components**		98
	8.9.1 Remote Latching Between Rotor and Rods		98
	8.9.2 Tag Bar-Stop Pin		98
	8.9.3 Rotor/Stator Position by Downhole Sensors		98
	8.9.4 Tag Bar Screen		98
	8.9.5 Spring Loaded Tag Bar		99
	8.9.6 No Go Tag on Top of the Stator		99
	8.9.7 Multi-Intake Sub		99
	8.9.8 Safety Valve and PCP's Pumps		99

Chapter 9
DRIVE HEAD

9.1	**Drive Head Principle**	101
	9.1.1 Drive Head Function	101
	9.1.2 Drive Head Types	101
	9.1.3 Rod String Twist and Back Spin Energy	102
	9.1.4 Drive Head Brake and Safety Protection if Sudden Stop	105
9.2	**Drive Head system**	106
	9.2.1 Hydraulic Power of the Pump	106
	9.2.2 Direct Belt Drive Head	108
	9.2.3 Mechanical Belt and Gear Reducer Drive Head System	108
	9.2.4 Variable Speed Drive Head (VSD)	108
	9.2.5 Mechanical Belt Drive System	109
	9.2.6 Hydraulic Drive System	109
	9.2.7 Direct Drive Head and Permanent Magnet Motor	109
	9.2.8 The Electronic Variable Speed Drive (VSD)	110
9.3	**Drive Head Code ISO Norm**	111
9.4	**Drive Head from Manufacturers**	111
	9.4.1 Belt or Gear Electric Standard Drive Head	111
	9.4.2 Gear Drive Head with Hydraulic or Electric Motors	112
	9.4.3 Direct Drive Head with Permanent Magnet Motor	115
	9.4.4 Drive Head PCP's Manufacturers Model Identification	116
9.5	**Rotary Seal Stuffing Box**	117
9.6	**Rod Lock Bop**	118
9.7	**Tubing Rotator**	119

Chapter 10
INSTALLATION, OPERATION AND MAINTENANCE

10.1	**Installation**	121
	10.1.1 General Consideration	121
	10.1.2 Pre-operational Checks	121
	10.1.3 Running in Stator and Tubings	122
	10.1.4 Running in the Rotor and the Rods String	122
	10.1.5 Setting up of the Drive Head and the Motorized Driving System	125

10.2	Start-up	126
	10.2.1 Pre-start Check	126
	10.2.2 Operating Procedures	126
10.3	Intervention Time	126
10.4	Operating Maintenance	127

Chapter 11
INSERT PCP SYSTEMS

11.1	An Economical Completion	129
11.2	PCP Insert Components and Operating Procedure	129
11.3	Pump Insert Advantages	132
11.4	Pump Insert Manufacturers	132

Chapter 12
ELECTRICAL SUBMERSIBLE PCP

12.1	Historic	133
12.2	ES-PCP	134
	12.2.1 ES-PCP Principle	134
	12.2.2 ES-PCP Components	135
12.3	Permanent Magnet Motors: New	139
	12.3.1 ES Permanent Magnet Motor Technology	139
	12.3.2 PMM Advantages and Benefits	139
12.4	ES-PCP Retrivable Through Tubing	140
	12.4.1 PCP's Component Retrievable Through Tubing	141
	12.4.2 ES Motor and PCP Retrievable Through Tubing	141
	12.4.3 Advantages of Through Tubing Conveyed PCPs or ES-PCPs	142
12.5	ES-PCP Advantages, Benefits and Constraints	143
12.6	ES-PCP's Applications	144
12.7	ES-PCP's Manufacturers	144

Chapter 13
NEW FEATURES ON PCP PUMP COMPONENTS

13.1	PCP High Capacities	147
13.2	PCP Multiphases Fluids	148
13.3	PCP and Thermal Heavy Oil Recovery	150
13.4	PCP High Temperature Elastomer Stator	151
13.5	PCP with Metal Stator	151
	13.5.1 The Metal Stator PCP Principle	151

		13.5.2 Metal Stator: M-PCP Vulcain™	152
		13.5.3 Others Metal Stator PCP Manufacturers	153
		13.5.4 PCP Metal Stator Patents	153
		13.5.5 PCP Metal Stator Benefits	154
13.6	PCP Uniform Thickness Stator Elastomer		155
13.7	PCP with Hollow Rotor		156
		13.7.1 PCP with Injection of Diluents in the Drain Producer	156
		13.7.2 Adapted Drive Head for Injection Through Hollow Rods	157
13.8	PCP with Metal or Ceramic Coated Rotor		158
		13.8.1 Spray Metal Coating Applied on PCP Rotors	158
		13.8.2 Ceramic Coating Applied on PCP Rotors	159
13.9	Variable Speed Drive and PCP		160
		13.9.1 Why a Variable Speed to Drive PCPs?	160
		13.9.2 Electric Variable Speed Drive PCPs	160
		13.9.3 Hydraulic variable speed drive	160
13.10	Downhole "Hydraulic Motor" to drive PCP		161
		13.10.1 The Hydraulic Lines	161
		13.10.2 Advantages of the PCP Hydraulic Drive	162
		13.10.3 Compared to ES-PCP	162

Chapter 14
PCP: MANY USES

14.1	PCP and Heavy Oil Process		163
		14.1.1 PCP and Thermal Production	163
		14.1.2 PCP and Cold Production	164
14.2	PCP and Water flooding		166
14.3	PCP for Dewatering Gas Wells		166

Chapter 15
MONITORING & CONTROLLER WELL PRODUCTION

15.1	Overview		167
15.2	Limitations, Constraints, Benefits		167
		15.2.1 Why Monitor Wells	167
		15.2.2 Benefits	168
		15.2.3 Parameters to be Checked	168
15.3	Monitoring & Production Optimization		169
		15.3.1 Production Monitoring and Real-time Flow Rate Model	169
		15.3.2 Data Communications	169
		15.3.3 Data Capture, Data Logger and Historical Matching Production	170
15.4	Monitoring Equipment		170
		15.4.1 Downhole Monitoring	170
		15.4.2 Surface Monitoring	171

15.5	**Well Management Software**	172
	15.5.1 Pump Manufacturers Well Management Software	172
	15.5.2 Well Management Benefits	174
15.6	**Monitoring Well Management Conclusions.**	175

Chapter 16
PCP SOFTWARE

16.1	**C-FER "PC-PUMP®" Software**	177
	16.1.1 PC-PUMP® Software Modules	177

Chapter 17
PUMP FAILURES & OPERATING PROBLEMS

17.1	**PCP Failures – Identification and Description**	181
	17.1.1 Rotor Failures	181
	17.1.2 Stator Elastomer Failures	182
17.2	**Operating Problems**	184
	17.2.1 No Flow Upon Start-up	184
	17.2.2 Less Flow	184
	17.2.3 Excessive Power Consumption, Higher Torque	184
17.3	**Surface Mechanical Problems**	185
	17.3.1 Stuffing Box	185
	17.3.2 Bearing on Drive Shaft	185
	17.3.3 Vibration Drive Head	186

Chapter 18
PCP ADVANTAGES AND LIMITATIONS

18.1	**PCP Artificial Lift Advantages**	187
18.2	**PCP's Actual Limitations**	188
18.3	**PCP Equipment Advantages**	188
18.4	**PCP' Well Monitoring and Data Transmission**	189
18.5	**PCP Well Management and Software**	189

Chapter 19
UNITS, PUMP PARAMETERS, NOMENCLATURE

19.1	**Conversion factors**	191
19.2	**PCP Parameters**	193
19.3	**PCP's Useful Formulas**	195
19.4	**Nomenclature**	197

Chapter 20
WEB SITES AND MANUFACTURERS

Chapter 21
PCP NORMALIZATION ISO 15136 -1:2001, -2:2006, -1:2010

21.1	PCP ISO Normalization Dated 2001-2006-2010..................	203
	21.1.1 Stator Code	205
	21.1.2 Rotor Code..................................	205

Bibliography	209
List of figures	213
Index	217

CHAPTER 1

What is a Progressing Cavity Pump? What is it Used for? What New Features Does it Have?

Progressing Cavity Pumps (PCP) are a type of positive displacement pump, also known as eccentric screw pumps or even just cavity pumps.

A positive displacement pump forces a trapped fixed volume of fluid to move, forcing or "displacing" the fluid from the suction to the discharge tubing. PCP are "constant flow machines", each rotor rotation will produce the same flow at a given speed (RPM) no matter what the discharge pressure. The PCP steel rotor pump rotates within a rubber stator, and the chambers formed by the intermeshing of the rotor and stator provide an effective displacement mixture of fluids, gas and solid into the pumping discharge.

Progressing Cavity Pumps and system (PCPs) are widely used in many industrial applications to transfer and dose many types of fluid, liquids and slurry. The sectors in and applications for which they are used include food, chemicals, paper plants, water treatment and dosing, mechanical engineering pump systems, environmental waste water pumps and new energies.

The reputation of Progressing Cavity Pumps (PCPs) is now well established in land and offshore oilfield applications for artificial downhole pump lifts for heavy, medium or light oil with gas, water and sand contents, coal bed methane and high volume fluids for dewatering and surface oil pumps transfer.

A Progressing Cavity Pump's design and the manufacturing process used for its rotor/stator fit and geometry and elastomer stator range have to be optimized for use with the well geometry and the fluid characteristics at downhole conditions.

New are the use of PCPs to pump fluid in more difficult oil well environments such as:
- High temperature fluids from thermal heavy oil recovery techniques such as Steam Assisted Gravity Drainage (SAGD) or Cyclic Steam Stimulation (CSS) or Continuous Steam Injection (CSI).
- Metal stator and specific elastomer stator designed for high temperature fluids.
- Multiphase fluids with high gas content with hydraulic regulator rotor/stator.
- Surface drive PCP with permanent magnet electric motors (PM).

- Downhole drive PCP with downhole permanent magnet electric motors (PM-PCP) or downhole hydraulic motors (HM-PCP).
- Downhole surface sensors for oilfield monitoring.
- Data transmission from wells for oilfield controls, improved production, increased life span pumps, reduced failures, energy and cost saving.
- PCP software for better pump selection and an improved design process, rotor and stator materials selection for application compatibility with the downhole fluid, temperature and environment and expected long run life.

1.1 INTRODUCTION TO THE PROGRESSING CAVITY PUMP

Inventor René Moineau: a French success story, "un peu d'histoire"

The Progressing Cavity Pump was invented by a French engineer, René Joseph Louis Moineau.

René Moineau, born in 1887, had a superb intellect and an inventive mind. He earned his pilot's licence (n° 554) and took part in an international seaplane competition at Deauville on August 31 1913. Taking the controls of "Breguet n° 8", he won the speed prize for 100 miles, at a speed of 99.6 km/h.

Soon afterwards, he decided to manufacture a high-performance compressor and became a graduate Doctor of the Faculty of Science at the University of Paris, having submitted a doctoral thesis on "the new capsulism". His research led him to apply for his first patent, application filed under number FR 695 539 on May 13 1930, and US 1.892.217 patented Dec. 27, 1932 relating to a new system, usable as a pump, motor, or as a simple transmission device. In the same year, he displayed a cardboard version of his superb new Moineau pump at the "Invention Exhibition".

René Moineau's thesis – "the new capsulism" – is available on the PCM web site.

Many other patents to do with horizontal progressing cavity pump concepts were filed as FR 736.434 on April 29 1932, FR 780.791 on January 26 1934, FR 787.711 on March 21 1935, FR 818.242 on May 26 1936, FR 997.957 on September 13 1945, FR 1.003.216 on December 17 and patents in many others countries including the UK and the US.

In 1932, helped by Robert Bienaimé from Gevelot Company, René Moineau then formed a new company – "Pompe Compresseur Mécanique" (PCM) – of which he became the Founding chairman. A small workshop located in Vanves employing about thirty people was built.

Figure 1

René Moineau (1887-1948), certificate N° 554, pilot of Breguet N° 8, 1913 Deauville competition.
Source: *Musée de l'Air et de l'Espace*.

Figure 2

René Moineau's "Capsulism" pump.

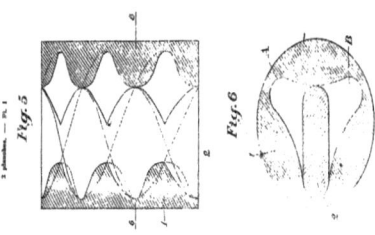
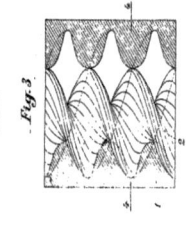
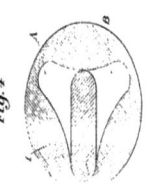
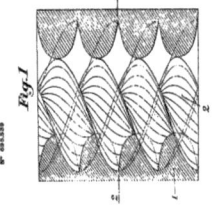

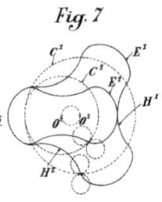
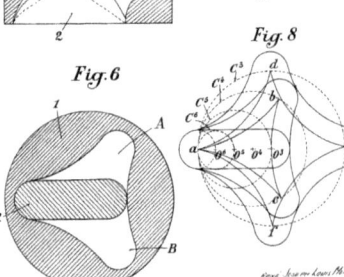

In 1936, René Moineau entered into a series of licensing agreements with foreign companies: Mono Pumps Ltd in England, Robbins & Myers in the US and Netzsch in Germany.

René Moineau continued to develop his pumping process right up until his death in 1948. His last invention was a multiple pump, based on the same principle, but useful when high flow rates were needed.

Figure 3

More recently dated 31/10/2006 a book on René Moineau's life 1887-1948 aviator and inventor.
Written by Jean Louis Moineau, his son.
Source: *Edition de l'Officine*.

1.2 MOINEAU: THE PUMP SUCCESS STORY FOR THE OIL INDUSTRY

One of the advantages of the Moineau pump is the ability to pump thick products without spoilage. This enables various uses as a surface pump in several industries, including chemical, food, coal, mining, textiles, tobacco and waste water treatment. It is also widely used as a surface transfer pump.

During the fifties, the Moineau principle was applied to hydraulic motor applications by reversing the pump function. The rotor/stator set driven under pressure by the drill fluid became an application of the Moineau principle widely used in the petroleum drilling industry to drive the drilling bits down to bottom hole.

From 1957, the Russians looked into the opportunity to use the Moineau system for drilling and petroleum production. In the seventies, they used Moineau pumps for the *"in situ"* production of heavy oils. They are rotated by a quadripolar electric submersible motor, running at 1450 rpm at 50 Hz.

In 1985, the partnership between the "Institut Français du Pétrole" and PCM, in a well operated by Total, ran its first Moineau pump driven from the surface by sucker rods. It was a simple, economical and rugged fitting suitable for all environments. It was suitable for use with heavy oils containing gas. This production method is having more and more success in countries with large viscous oil fields such as the US, Canada, Venezuela, Russia, Albania and China. It has also been very effective with the production of light and abrasive oils, and today it can produce high liquid flow rates.

PCM manufactures and sells PCP for downhole oil well production and for the processing and surface transfer industry all over the world. The European Headquarters is at PCM SA, 6 boulevard Bineau, 92300 Levallois-Perret, France, Tel: +33 (0)1.77.68.31.00, website: www.pcm-pompe.fr, www.pcm.eu. The French production unit is at Champtocé-sur-Loire, rue René Moineau.

In 1951, the German Group Netzsch acquired the Progressing Cavity Pump license to manufacture and distribute Nemo® Progressing Cavity Pumps according to the Moineau pump system. Nemo® took its name from: NEtzsch + MOineau.

The Canadian Company Kudu Industries was founded in 1989 by Robert and Ray Mills and is based in Calgary. Kudu manufactures and sells PCP pumps for oil well production all over the world.

The US Company Robbins & Myers, Inc. is a leading supplier of engineered equipment and systems for the industrial, chemical and pharmaceutical markets. Its subsidiary Moyno Inc is a manufacturer and marketer of Moyno® progressing cavity pumps. The trademark "Moyno®" sounds similar to "René Moineau", the inventor's name.

Other industrials based in Germany, Russia, China, Brazil and Canada will manufacture this type of pump and motor.

The Anglo-American manufactures refer to this pump as a *Progressing* (or *progressive*) *Cavity Pump* in the abbreviated form PCP. The cavities have the same volume, but the pressure *progresses* from suction to up to when the contents are discharged.

1.3 PCP BASIC PRINCIPLE

Progressing cavity Pumps are made of two elements: a metallic rotor and an elastomer stator. When the rotor turns inside the stator, the fluid moves along the pump axis inside the closed cavities that are between the rotor and the stator. The PCP flow rate is non-pulsating and constant. It depends on the rotor diameter, the pump eccentricity and the length of the stator pitch and is proportional to the rpm rotor rotation. The lift capacity or head rating of the PCP pump is dependent on the number of stator cavities, i.e. the total number of stator pitches.

The variation between these different parameters is catered for by several pump models which are available on the present market.

Each distributor develops its own product line by varying the geometry, the pressure and flow rate, as well as the elastomer class and the rotor surface coating. Many factors must be considered when choosing a pump model: the flow rate with an optimal rotation speed, head rating, casing diameter, presence of sand, suction conditions, torque and power limits and compatibility between the fluid pumped and the elastomer selection. In many cases, the result achieved in the field is the best method for choosing a pump model. But in the case of a new application, the operator needs to find the optimum pump characteristics, with guarantees for reliable running.

To ensure trust between users and the various suppliers, a new work item ISO Standard has been proposed and adopted. Its number is ISO 15136-1. It defines the full equipment, the user conditions, the model codification and the qualification tests.

The success of a petroleum production tool depends on its technical integrity and on the manufacturers', distributors', technicians', field development managers' and operators' abilities. The purpose of this technical book is to provide the criteria for selecting a progressing cavity pump and operational conditions. It is intended to provide sufficient and necessary information about these criteria.

CHAPTER 2

Progressing Cavity Pump: Operating Principle

A Progressing Cavity Pump is an original rotating pump. Its principle and its theoretical elements are described below.

2.1 THE PCP PRINCIPLE

The Progressing Cavity Pump (PCP) is a rotary positive displacement pump consisting of two components, a rotor rotating inside the stator.

For a single lobe PCP type 1-2, a single external helical rotor rotates eccentrically inside an internal double helical stator of the same minor diameter and twice the pitch length.

The rotor is a metallic rod machined into a single helical profile. The stator is an elastomeric material permanently bonded by injection process inside an end threaded steel tube support. The elastomeric stator form a double helical internal profile with a pitch double that of the rotor. The pitch length is the length of a 360° rotation of the crest trace of one of the helix lobes. Because the number of rotor and stator lobes differs by one, a fluid-filled cavity is formed between rotor and stator. For example PCP 1-2 is one lobe rotor and two lobes stator. For a multilobe pump, for example, the PCP 2-3 is two lobes rotor and three lobes stator. There is positive seal or fit or compression contact line between the metallic rotor and the elastomeric stator. This line of contact is called the seal line. One completely sealed *cavity* is a closed volume located between the single helical rotor and the double helical internal stator for a 360° rotation of the stator helix. Each independent sealed cavity will generate a given amount of pressure. The increase of the stator and rotor length will increase the number of *progressing* sealed *cavities* from suction to discharge and will increase the pump's pressure capabilities.

The rotor rotates at constant rpm inside the stator, one cavity is open as the other is closing so the PCP's flow rate is non-pulsating and constant; it depends on the cavity volume and the rotor rpm rotation.

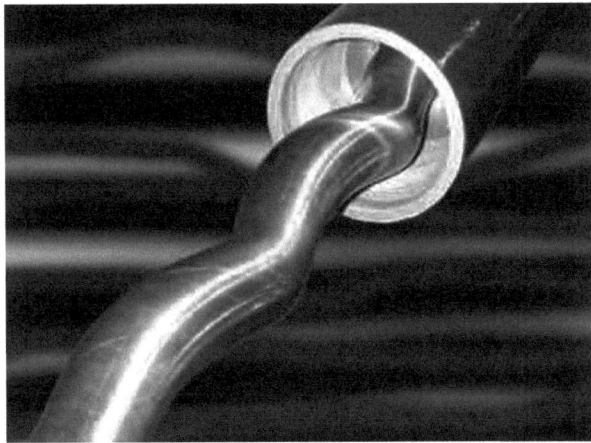

Figure 4

Single lobe PCP type 1-2, rotor and stator view.
Source: *National Oil Well*.

2.2 PCP GEOMETRY

A Progressing Cavity Pump (PCP) is a rotary pump essentially made up of two helical gears, the gear rotor rotating inside the gear stator.

The PCP's rotor and stator are made differently from those of screws and nuts. The centre of a screw always stays lined up with the centre of a matching nut, but the centre of a PCP rotor changes its position along its axis.

The rotor rotates around his longitudinal axis which is parallel but spaced to the stator axis. The stator in the external gear has one more thread or tooth or lobe than the internal rotor element. Whatever the different number of threads or lobes between the two elements, they must always be differentiated by one unit.

For a single lobe PCP type 1-2 a single external helical rotor rotates eccentrically; it is rolling with no sliding inside an internal double helical stator of the same minor diameter and twice the pitch length.

The helical winding of the profiles around their rotating axis creates, between the two helical elements, volumes whose length is equal to the stator pitch. The rotor turns round inside the stator element, the volume of fluid-filled cavity or enclosed cavity formed between a rotor and a stator will move or displace without deformation according to a helical motion along the stator element. The pump will enable a discharge under pressure or a fluid expansion, with no necessity to use a check valve. The pressure will increase only after the first rotor rotation; this motion leads to the formation of "closed cavities", delimited by the rotor and the stator which move axially from suction to discharge.

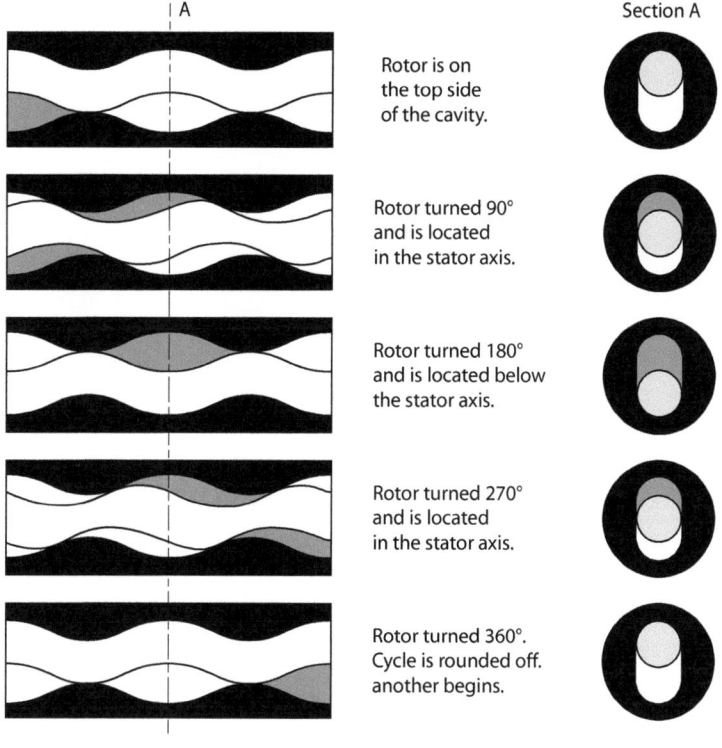

Figure 5-1

Operating principle diagram of the PCP 1-2.
Source: *PCM*.

Using this principle, a rotating positive displacement pump is achieved which:
- Has no check valve or non-return valve.
- Provides a uniform flow rate at a constant rpm stator rotation without any pulsation or jerk.
- Is capable of conveying very liquid to very viscous fluids, even when containing solids and gas.

Figure 5-1 is a principle scheme where the stator (shown in longitudinal section) is fixed. The rotor (pitch on right) turns clockwise. Cavities are formed between the two gears, which are open at the left extremity, when the rotor turns. The isolated cavities expand, separate and come out at the other extremity, then decrease slowly and disappear.

Figure 5-2 is another presentation of the operating principle showing for each section the stator shape, the rotor position inside the stator, and the movement of oil running into a cavity.

So, the fluid moves from left to right. The discharge and the suction are always isolated from each other by a seal line of a constant length.

If the rotor runs in reverse gear, the closed cavities move from right to left according to the same principle. Therefore the pump is definitely reversible.

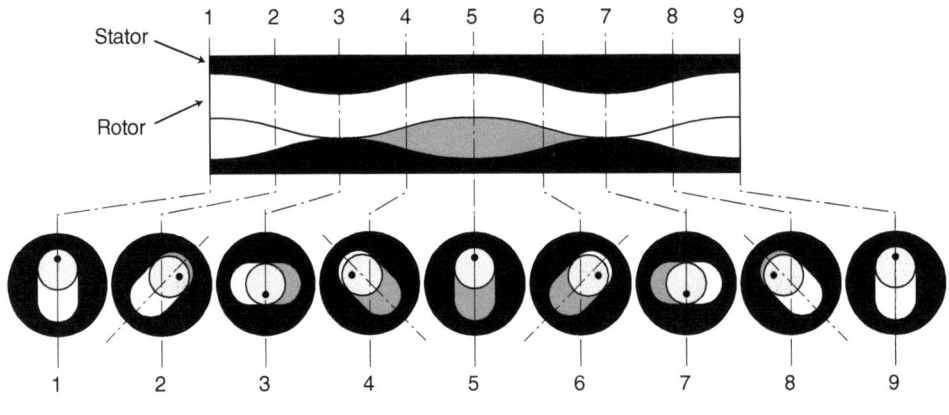

Figure 5-2

PCP 1-2 operating principle.
Source: *Petrobras*.

2.3 ROTOR-STATOR GEOMETRIES

The profile geometries of a rotor and a stator are made differently.

2.3.1 Theoretical Elements

Three conditions are necessary in order to obtain "closed cavities":

1. The rotor must have one tooth (lobe) less than the stator.

2. Every tooth (lobe) of the rotor must always be in contact with the inner surface of the stator.

3. The rotor and the stator, as they are defined above, longitudinally constitute two helical gears.

• **First condition:** *The rotor must have one tooth (lobe) less than the stator.*

The hypocycloid H_1 (stator) with n teeth, whose base is the circle C_1 (O_1 R_1), is connected to the hypocycloid H_2 (rotor) with $(n-1)$ teeth whose base is the circle C_2 (O_2, R_2) by the relation: $R_2/R_1 = (n-1)/n$. These two curves provide two gears, one inside the other (Fig. 6-1).

H_1 being fixed, when H_2 turns in a certain direction, its centre O_2 draws in the opposite direction a circle of centre O_1 and of radius O_1O_2 such that:

– Stator central line O_1 axis.
– Rotor central line O_2 axis.

$$\text{Distance } O_1O_2 = E = \text{eccentricity of the pump}$$

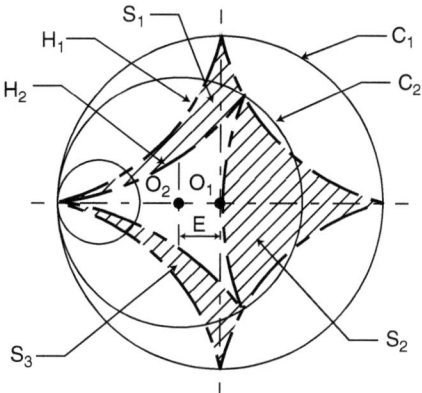

Figure 6-1

Pump type 3-4: Hypocycloids H_1 and H_2.
Source: *PCM*.

• **Second condition:** *Every tooth (lobe) of the rotor must always be in contact with the inner surface of the stator.*

During this motion, the vertexes (summits) of H_2 are always in contact, the contact lines, with H_1 and on these curves appear closed surfaces S_1, S_2, S_3 of variable area, whose sum $(S_1 + S_2 + S_3)$ remains constant.

If we replace H_1 and H_2 by their envelopes E_1 and E_2 (Fig. 6-2) of an identical circle C of any diameter D which the centre would describe H_1 and H_2, the previous characteristics remain unchanged and the first condition is thus satisfied.

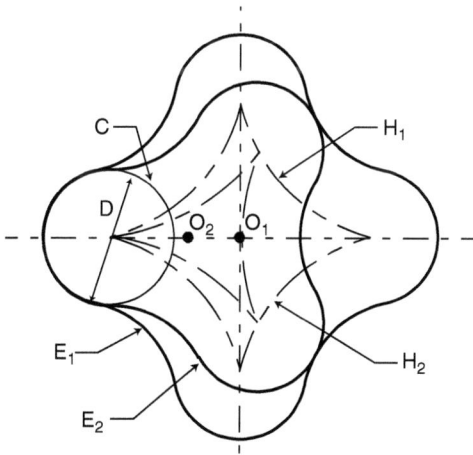

Figure 6-2

Pump type 3-4: Hypocycloid envelopes.
Source: *PCM*.

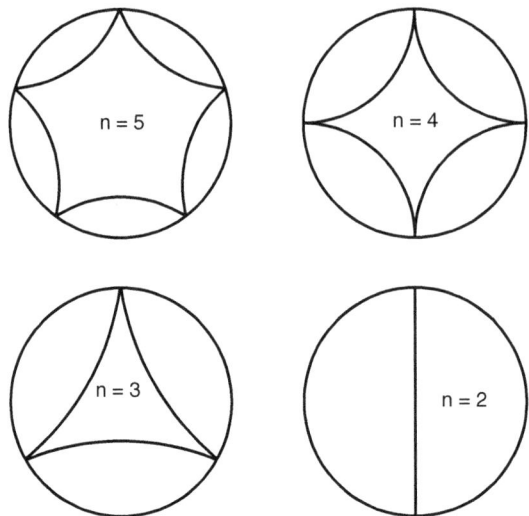

Figure 6-3

Hypocycloids profile cross sections of helical elements from various types 1-2 ($n = 2$), 2-3 ($n = 3$), 3-4 ($n = 4$), 5-5 ($n = 5$).

The geometrical motion is based on the concept of a hypocycloid which defines a hypocycloid as the curve traced by a point on the circumference of a rolling circle inside a fixed circle that is twice the diameter.

The type 1-2 hypocycloid path is always a straight line.

• **Third condition:** *The rotor and the stator, as they have been defined above, longitudinally constitute two helical gears.*

To satisfy the second condition, the profiles E_1 and E_2 move longitudinally as helices whose pitch ratio corresponds to the ratio of the number of teeth.

The helical winding of the surfaces S_1, S_2, S_3 makes it possible to obtain, between the two helical elements, "closed cavities" whose length is equal to the pitch of the external element.

To achieve total sealing of the pump, the engagement length of the rotor in the stator has to be at least equal to one pitch of the stator.

The portion of the contact lines of the rotor and the stator between the high-pressure cavities and the low-pressure cavities, constitutes the leakage line.

2.4 MONO AND MULTI-LOBES PUMP GEOMETRIES

Progressing Cavity Pumps consist of two elements: the rotor and the stator.

The rotor and stator axes of a progressive cavity device are eccentric to each other and the rotor rolls with no sliding inside the stator cavity.

The geometry of the assembly is such that it achieves two or more series of separated cavities. When the rotor turns inside the stator, the cavities move in a spiral from one end of the stator to the other, creating the pumping action. The driving system makes it rotates on itself. When the rotor has completed one revolution, its axis has also achieved a rotation in the reverse direction around the stator axis, but remains parallel. The *hypocycloid* trajectory of a rotor centreline around the stator centreline generates a nutation rotation movement in the direction opposite to the rotor rotation. Because the number of rotor and stator lobes differs by one, a fluid-filled enclosed cavity is formed between a rotor and a stator. During the rotor revolution the nutation movement of the rotor centreline to the stator centreline causes the enclosed cavity to be displaced along the pump axis from inlet to outlet.

For multilobe units, the same principle applies, except that there are *multiple* cavities created, which all move along the axis of the stator during the rotor revolution.

The geometry of pumps ($L_r - L_s$) is generally defined by two numbers L_r and L_s. The first (L_r) being the number of lobes of the rotor, and the second (L_s) being the number of lobes of the stator.

2.4.1 Mono-Lobe (1-2 Pump) Geometry

The geometry of **a mono-lobe pump** with a single helical rotor and a double helical stator is described as a "1-2 pump" with $L_r = 1$ and $L_s = 2$ (Fig. 7, 8).

The mono-lobe pump "1-2" geometry is made up of a helical gear with two helices, one inside the other:

- The metal rotor, the internal one, is a simple helix determined by the number 1.
- The stator, the external one, is a double helix with twice the pitch length of the rotor determined by the number 2.

The rotor is not concentric with the stator. Therefore, the motion of the rotor inside the stator is a combination of two motions:

- A motion rotation around its own rotor centreline axis in one direction.
- A motion rotation in the opposite direction of its own centreline around the centreline axis of the stator.

The geometry section of the helical gear formed by the rotor and the stator is fully defined in Figure 7.

2.4.1.1 Mono-Lobe Diameters and Eccentricity: E and D

Referring Figures 7:
- The eccentricity is the distance between the centreline of the rotor O_2 and the centreline of the stator O_1 and is symbolized as: E
  - O_1 centreline of the stator axis
  - O_2 centreline of the rotor axis
  $$E = O_1O_2 = \text{eccentricity of the pump}$$

- The thickness of a single helical rotor (minor rotor diameter) is symbolized as: **D**.
- The diameter of the helix rotor (major diameter) is symbolized as: **(D + 2E)**.
- The width of the double-threaded helix in the stator is: **D** (minor diameter of the helix stator) and **(D + 4E)** major diameter of the helix stator.

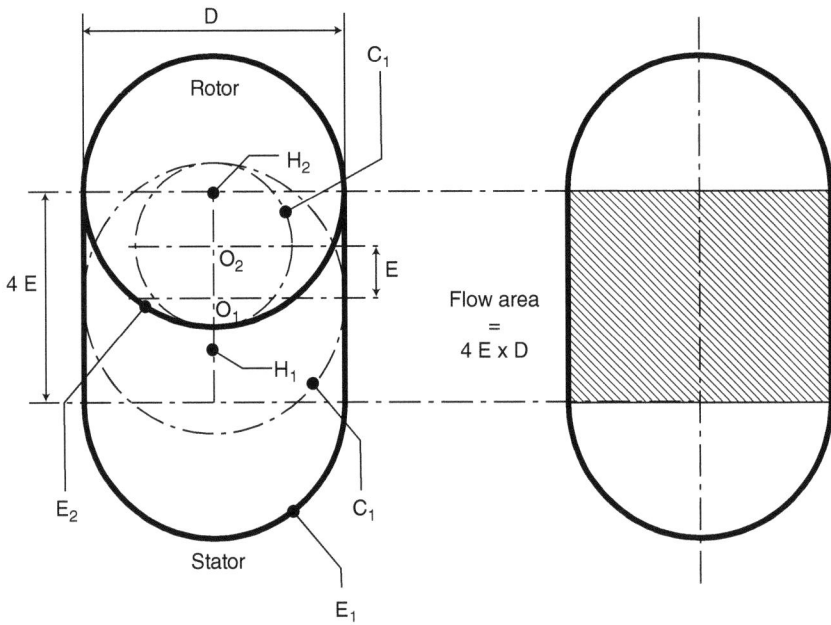

Figure 7

Pump 1-2 geometry.
Source: *PCM and IFP Energies nouvelles*.

2.4.1.2 Mono-Lobe Pitch Length: P

The pitch length is defined as a length of 360° rotation of the crest trace of one of the helix lobes, and is symbolized as **P**, shown in Figure 8.

Therefore, the pitch lengths of the rotor and the stator are more clearly symbolized as:

P_r pitch length of the rotor

P_s pitch length of the stator

For a mono-lobe 1-2 pump: $P_s = 2 P_r$

The pump kinematics i, is equal to the ratio between the pitch lengths P_r of the rotor and P_s of the stator or equal to the ratio between L_r lobe rotor and L_s lobes stator: i.e.: 1/2.

$$i = L_r/L_s = P_r/P_s = 1/2$$

Chapter 2 • Progressing Cavity Pump: Operating Principle

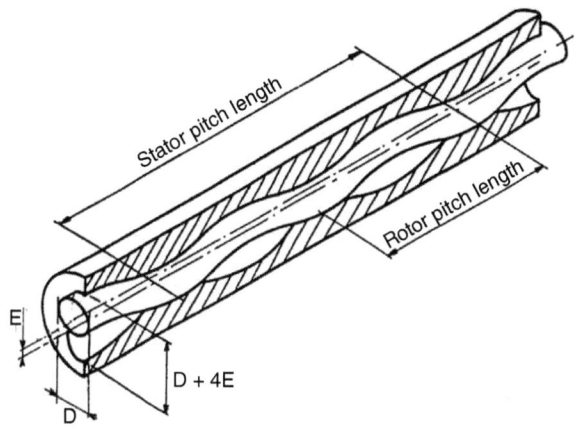

Figure 8-1
Rotor – Stator perspective.

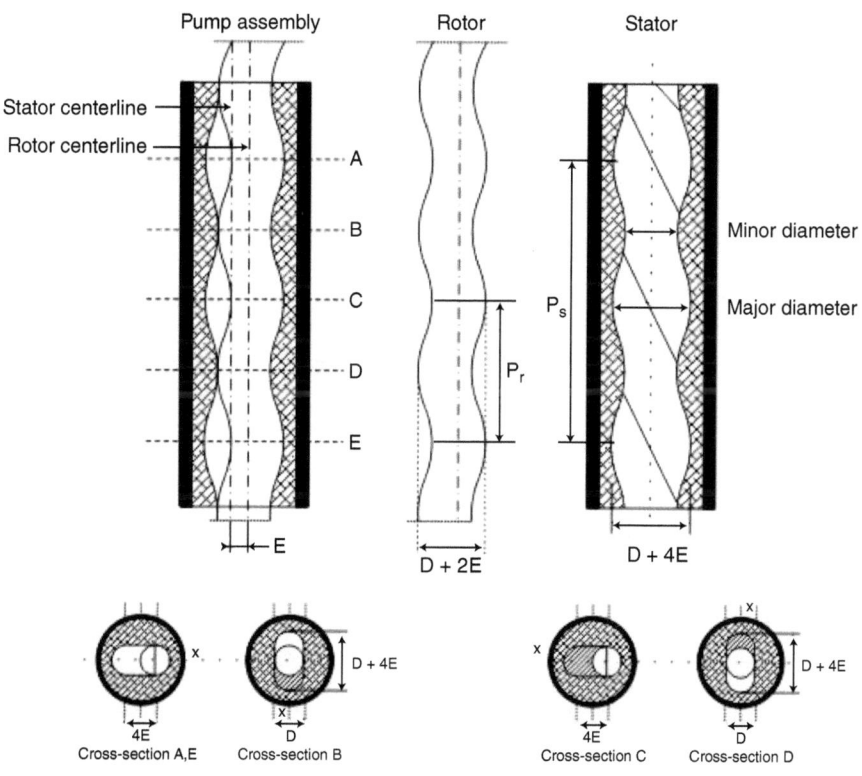

Figure 8-2
Mono lobe (1-2) Rotor and stator geometry.
Source: *IFP Energies nouvelles*.

2.4.1.3 Mono-Lobe Cavities

Cavities that are lenticular, spiral, and separate volumes are created between the stator and the rotor when they are assembled. The cavities are formed as a result of the additional helix in the stator cavity. Each closed cavity of P_r pitch rotor length moves in spiral around the axis of the stator, progressing from the inlet to the outlet as a consequence of the rotor rotation.

Along the stator axis, the stator section "turn" circumferentially until a complete 360-degree turnaround is reached at the other end to create a stator cavity length equal to the pitch length of the stator = P_s.

To create an enclosed cavity between rotor and stator, the rotor section completes its first 360-degree turnaround at the middle of the stator, and the second 360 degree turnaround during the second half of the stator pitch.

The enclosed cavity length created between stator and rotor is the axial distance of the rotor pitch (P_r) equal to one-half of the stator pitch (P_s) for a 1-2 pump: $P_r = P_s/2$.

The pump entrance's first half cavity and the last half cavity are not used to gain pressure.

The number C of effective working enclosed cavities for a mono-lobe 1-2 pump effective ($L_r = 1$) is:

$$C = \frac{H_s}{P_r} - 1$$

where:
- H_s is the total length of the stator.
- P_r is the length of the rotor pitch.

2.4.1.4 Description of the Rotor Pump 1-2

The rotor is made from high-strength steel and coated with a resistant substance (chromium plating) so as to minimize the abrasion and to decrease the coefficient of rotor/stator friction.

The rotor of helical shape and of circular section is defined by:
- Diameter (minor helix rotor diameter) D
- Rotor/stator eccentricity E
- Major helix rotor diameter $D + 2E$
- Pitch of the rotor helix mono-lobe $P_r = P_s/2$ (pump 1-2)

2.4.1.5 Description of the Stator Pump 1-2

The stator is made from an elastomer that is particularly designed to withstand petroleum effluent (crude, salt water, gas), and downhole temperature.

The stator of helical inner shape is defined by:
- Minimum width of the helix stator section D
- Maximum width of the helix stator section $D + 4E$
- Pitch of the stator helix $P_s = 2 P_r$ (pump 1-2)

2.4.2 Multi-Lobes Pump Geometry

A multi-lobe pump is made of one rotor with the L_r number helices set one into the other, and inserted into a stator including $L_s = L_r + 1$ number of helices. In cross-section the rotor and the stator have wave-shaped outline profiles, each helix correspond to a "lobe".

The pumps are calling "$L_r - L_s$ pump". For example, a pump having a rotor including 4 lobes and a stator including 5 lobes is called "4-5 pump".

For a ($L_r - L_s$) multi-lobes pump

L_r: number of rotor lobes

L_s: number of stator lobes

$$L_s = L_r + 1$$

The pitch length for a multi-lobe pump a is defined as a length of 360° rotation of the crest trace of one of the helix lobes,

P_r: pitch length of the rotor helix multi-lobes $P_r = \dfrac{L_r}{L_r + 1} \times P_s$

P_s: pitch length of the stator helix multi-lobe $P_s = \dfrac{L_r + 1}{L_r} \times P_r$

The pump kinematics i is the ratio between number of rotor lobes (L_r) and stator lobes (L_s) and is equal to the ratio between pitch lengths P_r of the rotor and P_s of the stator.

$$i = L_r / L_s = P_r / P_s$$

$P_r = 2/3 \, P_s$ (for 2-3).

$P_s = 3/2 \, P_r$ (for 2-3).

The number C of effectives enclosed cavities for a multi-lobes ($L_r - L_s$) pump is:

$$C = L_r \left(\dfrac{H_s}{P_r} - 1 \right)$$

The total fluid displacement per one rotor revolution is equal to the product of one enclosed volume cavity times the number C of enclosed cavities.

2.4.2.1 Advantages of Multi-Lobe Pumps

Multi-lobe pumps compared with 1-2 pumps have, for the same number of lobes, the following advantages:
- An increase in displacement, equal to the number of rotor lobes L_r. So, according to the Figure 9, a 5-6 lobe pump has a flow rate capacity nearly 3 times higher than a 1-2 pump.

- Lower value of the eccentricity E which decreases the rotor unbalance, the pump vibrations, and the dimensions of the eventual connection to the drive shaft, if the pump is connected with a submersible electric motor.
- A lower head rating by cavity, generated by an increased number of contact lines separating cavities. Consequently, for the same pump length, it is possible to increase the total head rating by augmenting the lobe numbers.
- For the same flow rate, a multi-lobe pump will have a lower speed. So, according to Figure 9, between a 5-6 pump and a 1-2 pump, the rotating speed is reduced by nearly 3 times. On the other hand, the resistant torque is increased.
- The Standard 1-2 "S" geometry Netzsch pump compared to multi-lobe geometries extends the standard range by increasing the flow.
- The 2-3 "D" geometry delivers 145% of flow compared to a standard 1-2 pump's flow.
- The 1-2 "L" extended pitch geometry provides 200% of flow.
- The 2-3 "P" extended pitch geometry provide 285% of flow designed for use in water industry applications.

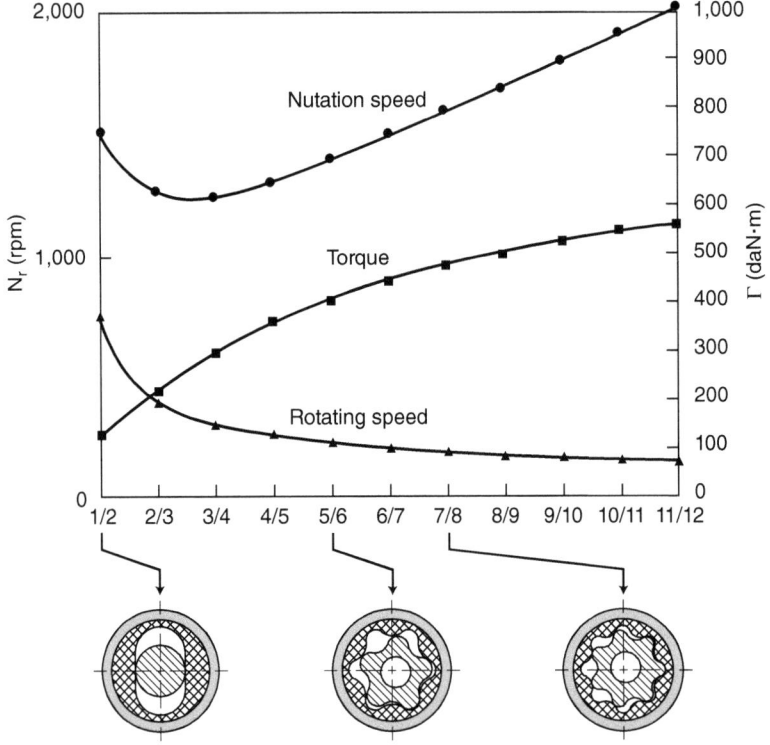

Figure 9

Multi lobe pump relationship between: lobe pattern i, rotating speed N_r, nutation speed, torque Γ for a flow rate of 4000 m³/d and $\Delta P = 1$ MPa (10 bar).

Source: *According to M.T. Gusman and D.F. Baldenko.*

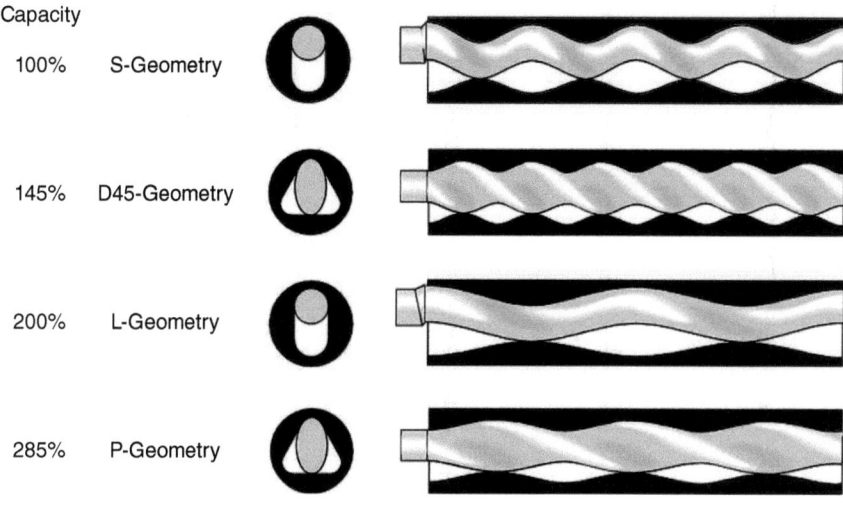

Figure 10
Multi lobe pumps geometry example.
Source: *Netzsch*.

2.4.3 Multi-Lobe Geometry Pump's Applications

Multi-lobe geometry pumps are mainly designed for use with surface pump fluid transfer in industry applications and downholes when large production flow rates are expected such as water wells and water flooding (chapters 11.2 and 12.2).

Multi-lobe systems are mainly used as downhole hydraulic motors, in petroleum drilling. They can withstand large flow rates at low rotating speeds, so they reach high output power and torque.

2.5 PUMP MANUFACTURING PROCESS

2.5.1 Rotor Manufacturing Process

The rotor is made from high-strength steel. The rotor manufacturing technologies process uses electronically controlled five-axis machines to cut the complex three-dimensional shapes of the PCP's rotors.

The rotor is then polished and coated with a resistant substance such as chromium plating to minimize abrasion and decrease the coefficient of rotor/stator friction.

The thickness of the chromium plating products determines the desired final diameter of the rotor to a desired stator fit. It depends too on the possible swelling of the elastomer generated by the pressure and the abrasive nature of the pumped products.

2.5.2 Stator Manufacturing Process

The stator core mandrel is machined as the rotor process. The PCP's stator is made by injection process. The injected material is a specific elastomer (e.g. hydrogenated nitrile (HNBR), fluorocarbon (FKM) Viton®)) depending on the downhole conditions in term of fluids and temperature.

The raw elastomeric material comes in strips that are fed into a heated auger in the injection machine, and is moved by the piston under high pressure to fill the voids between the stator steel tube support and a specific core stator mandrel that is positioned inside the tube and centred. The core stator is an exact duplicate of the stator "void" or "inverse image" that has dimensions corrected for elastomer shrinkage. During the injection process when heated at specific a temperature (180°C, for example), the elastomer undergoes vulcanization. The core stator is removed by traction on it, the stator then cools in the autoclave and during the cooling process, the elastomer shrinkage is determined by the elastomer's properties.

2.6 PUMP ROTATION TRANSMISSION

The rotor is driven into rotation either by a joint system or a flexible shaft, if the pump is connected to a submersible electric motor either by drive strings whose flexibility enables the nutation motion of the rotor axis in the opposite direction. The rod's drive string characteristics are described in chapter 8.

2.7 PCP AND OIL WELL APPLICATIONS

Generally, Progressing Cavity Pumps are driven from the surface by solid drive rods.

They are made of (Fig. 11):
- A **stator** screwed in at the tubing extremity.
- A **rotor** fixed at the solid rod's extremity or small diameter drive strings.
- A **drive head** adapted on the wellhead and taking in the efforts absorbed by the drive strings.
- An **electric motor** (or other: hydraulic, gas) with speed reducer or variable speed drive.

Other Systems

For low producing wells, a special system – the "Pump Insert" – has been designed which enables a sealed insert-pump anchor fixed on the tubing production extremity. This gives a reduction of operating time without running out the tubing. This production system is described in chapter 11.

Another production process relative to Progressing Cavity Pumps is the mounting of a submersible electric motor onto the pump. This is described in chapter 12.

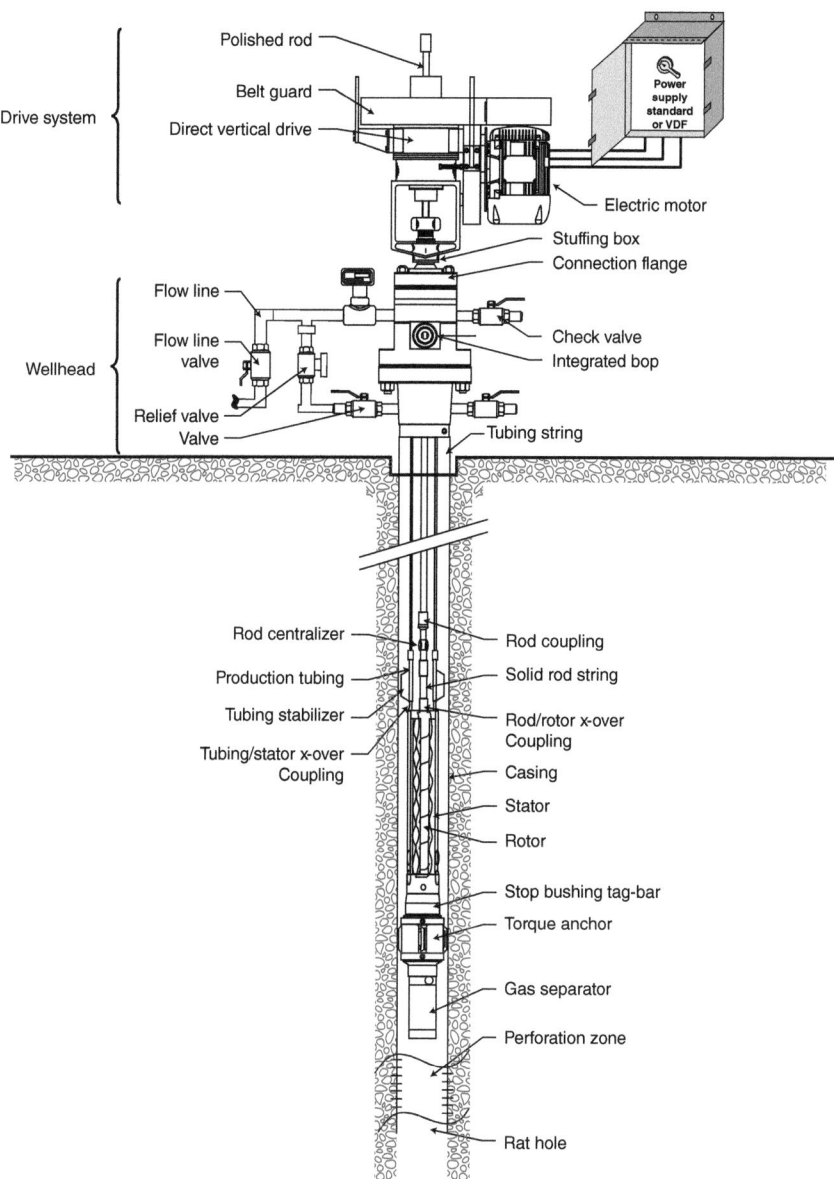

Figure 11

PCP typical configuration.
Source: *Protex*.

CHAPTER 3

PCP Characteristics

The majority of manufactured pumps essentially consist of a one-lobe rotor and a two-lobe stator. The following description relates to this "1-2 pump" type.

3.1 MONO-LOBE PUMP CHARACTERISTICS

3.1.1 Pump Kinematics

The rotor, which is the rotating internal component, is interdependent of the solid rod string run by a surface drive head. It is very precisely manufactured in high-strength steel, and chromium plated to minimize abrasion and friction between rotor and stator.

The stator, which is the external pump component, is interdependent of the production tubing and remains fixed or anchored during production. The stator is usually constituted of a steel tube and an injected elastomer moulded element, encased inside. The elastomer is injected and moulded between a core and the external tube, in the configuration of an internal double helix whose pitch is twice the rotor's. The elastomer is chosen in consideration of the chemical and physical environment of the well (see chapter 3-3).

To resolve the problems encountered with high temperature pumping (349°C-660°F), the possibility of using metallic stators is now offered by two pump manufacturers (see chapter 13-5).

The rotor/stator torque is mainly defined by its eccentricity, the rotor's diameter and the stator's pitch. The rotor rotations conduct to identical cavity formations, but are separated (Figs. 8-1 and 8-2). Each of these cavities is full of fluid, having for length the stator pitch, unwound around the rotor in transmitting the fluid to the further cavity. In each cross section of the pump, two cavities are constantly opposed to the rotor. The rotor rotation inside the stator is a combination of two motions:
– A rotor rotation to the right around its own axis.
– An eccentric reverse rotation around the stator axis.

The co-ordination of these motions creates a nutation of the rotor through a stator section (Figs. 5-2 and 7).

3.1.2 Pump Volume Displacement

When the pump is driven into rotation, the cavities move longitudinally from suction to discharge, creating a pumping action in the process. During each rotation, the fluid volume contained in the cavity is displaced by a length equal to the stator pitch. Because the cavity volume between the rotor and the stator remains constant, whatever its cross section, the pump enables a uniform, non-pulsating flow rate. The **displacement** is determined by the fluid volume produced in one revolution of the rotor. So, the function is symbolized as follows:

- E: Eccentricity rotor/stator.
- D: Rotor diameter.
- P_s: Stator pitch, cavity length.

P_s is also the cavity length and the distance of displacement of the cavity volume for one rotor rotation.

The net fluid cross-sectional area is constant along the axis of the pump.

$$\text{Flow area} = 4E \times D$$

The cavity volume V is equal to this fluid area between rotor and stator time the cavity length which is the stator pitch P_s of the 1-2 pump. Consequently, for one rotation, the pump cylinder is equal to (Figs. 7 and 8-2):

$$\text{Cavity volume} = V = 4E \times D \times P_s$$

So for a defined constant rotating speed, the flow rate is regular. This is a remarkable characteristic of the Progressing Cavity Pump.

As it is a positive displacement type pump and considering the efficiency for a given head rating, the pump flow rate is directly proportional to its cylinder capacity and its rotation speed. N being the number of rotations per minute (rpm), the calculated flow rate per minute Q_c is:

$$Q_c = 4E \times D \times P_s \times N$$

The actual pump flow rate Q_a is determined by considering a leak rate Q_s, which is:

$$Q_a = Q_c - Q_s$$

The flow rate depends on the volume of cavities and the rotor rotating speed. A limited speed along the helical way is recommended. It is fixed at around 10 m/s.

The manufacturers generally characterize their pumps with a reference to the number of rotations per minute for a daily flow rate. The standard ISO 15136-1 plans to codify the daily flow rate of the pumps at 100 rotations per minute (100 rpm).

3.1.3 Pump Head Rating

The differential pressure between suction and discharge generates a fluid leakage between the two successive cavities from high to low pressure, consequently a pressure gradient is delivered along the pump.

This pressure gradient depends on the pumped fluid characteristics. The liquids (quasi incompressible fluids) generate linear pressure gradients, whereas high gas content fluids have pressure gradients which are increasing quasi-exponentially from suction to discharge.

The pump head rating is determined by:

1. The number of cavities formed between the rotor and the stator.
2. The head rating developed into an elementary cavity, which depends on:
   - The light internal recirculation flow (slippage, leakage or fit) between rotor and stator (the diameter of the rotor is slightly bigger than the minor diameter of the stator).
   - The pumped fluid characteristics (higher head ratings are achieved with viscous fluids).
   - Values determining the geometric profile of the pump: diameter and rotor pitch, eccentricity.
   - The chemical composition of the elastomer.
   - The mechanical characteristics of the elastomer.
   - The thickness of the elastomer.

In a first approach, a reference value of head rating δp for a cavity of about 300 to 500 kPa (3 to 5 bar) may be chosen. To withstand a high head rating, the PCPs are made with a large number of cavities. Then, the total head rating is evaluated at:

$$\Delta P = \delta p(2n_p - 1)$$

where n_p is the number of stator pitches P_s,

The pressure is generated by each enclosed effective rotor cavity and the number of effective rotor cavities is equal to 2 n_p because $P_r = P_s/2$.

3.1.4 Pump Torque

The PCP are rotating pumps generating a torque to start the pump or starting torque and during the pumping operation an operational resistant torque or running torque.

3.1.4.1 Starting Torque

This is the initial torque necessary to start up the pump. It is quite often much higher than the operational resistant torque. Therefore, the surface motor and the drive strings must be able to drive up the initial starting torque. Also, a security coefficient should be considered.

This starting torque must be initially measured and noted.

3.1.4.2 Operational Running Torque

The rotor rotation transfers the fluid from one cavity to another, thus producing a differential pressure. The energy required to generate this action needs sufficient resistant torque for the rotor and drive strings. This torque depends on:

- The pump hydraulic power (directly proportional to the head rating).
- The chromium plating quality of the rotor.

- The class of elastomer, the lubricating characteristics of the pumped fluid.
- The length of the pump.

A practical relationship allows the evaluation of the maximum resistant torque:

$$\Gamma_{(daNm)} = 1.63 \times V_{(cm^3)} \times \Delta P_{(kPa)} \times 10^{-5} \times \rho^{-1}$$

V being the pump cylinder capacity of the pump and ρ its efficiency (for the evaluation, $\rho = 0.7$).

3.1.5 Load on Thrust Bearing Drive Head

The head rating generated by the pump produces a tensile stress on the drive string. This force should be withstood by the thrust bearing of the motor system placed into the wellhead.

The load on thrust bearing or tensile stress on the drive string has the following value:

$$F_b = \frac{\pi \times \Delta P \times (2E + D)^2}{4}$$

The major diameter rotor ($2E + D$) is then a characteristic value of the pump, giving a calculation basis to define the driving equipment characteristics.

3.2 MULTI-LOBE PUMP CHARACTERISTICS

3.2.1 Theoretical Rotating Speed

Let V_0 be the flow rate corresponding to one rotor rotation.

The pump rotor is simultaneously put through a rotating speed N_r around its centreline, and a speed N_n of nutation speed describing a cylinder of radius E. around the pump axis, in an opposite direction to N_r.

For an ideal pump without leakage, crossed over by a flow rate Q_0, we have:

$$N_r = \frac{Q_0}{V_0} \text{ (rpm)}$$

$$N_n = L_r \times N_r \text{ (rpm)}$$

If S_0 is the cavity cross-section perpendicular to the axis and the stator pitch length P_s, the contained volume into a pitch is $S_0 \times P_s$ and the flow rate by rotation of the rotor around its axis is:

$$V_0 = S_0 \times P_s \times L_r$$

Consequently:

$$N_n = \frac{Q_0}{S_0 \times P_s}$$

With equivalent pitch content a multi-lobe pump rotates L_r times more slowly than a cylinder rotor pump. This is a great advantage for the large flow rate pumps which run down in very inclined wells.

3.2.2 Flow Rate

The flow rate depends on the volume cavities and the rotor speed. A limited speed along the helical way is recommended, fixed around 10/s.

3.2.3 Pump Torque

The theoretical torque for a pump is:

$$\Gamma = \Gamma_0 \times \Delta P \times D \times E \times P_r$$

where:
- Γ is the torque corresponding to rotor pitch length.
- ΔP the head rating.
- E the eccentricity.
- D the reference diameter of the pump ($D = 2 \times E \times L_s$).
- P_r the rotor pitch length.
- Γ_0 the specific torque of the pump (about 2.5 daN.m for a pump of $D = 100$ mm).

3.3 ELASTOMERS STATOR CHARACTERISTICS

3.3.1 Elastomer Selection *versus* Pumped Fluids

The difficult part of the PCP is the stator manufacturing process with the specific moulded elastomer encased and bonded into a steel housing pipe.

Elastomer selection is a critical step in the PCP's design and selection as it influences the pump lifetime, performance and operating costs.

Elastomer stators are part of the ISO 15136-1 (2010-4). The stator is identified with "ISO-vvv-hhhh-eee" where "eee" is the elastomer code and the annex A "elastomer for requirements" and annex D "for testing and selection".

The physical characteristics of an elastomer can vary according to the fluid pumped and the *in situ* conditions. Consequently, different elastomer formulations and properties may be distinguished for:

1. Heavy and very abrasive oils (< 18°API).

2. Heavy and medium viscous oils, more or less abrasive, for a working temperature below 100°C.

3. Light oils more or less containing aromatics and carbon dioxide.

4. Pumped oil at high temperatures (references up to 160°C).

For a specific well, accurate well data is required to design a complete PCP's system. A good knowledge of downhole well parameters, such as completion, flow rate, fluid properties and chemical composition, well bore temperature such as:

- API gravity.
- Aromatics, CO_2, H_2S.
- Water-cut.
- Temperature at pump level.
- Abrasion level/sand content.
- Gas/liquid ratio.

Will determine the correct elastomer utilization criteria and its chemical formulation.

3.3.2 Measurements Characterising an Elastomer

An elastomer is physically characterized by:
- Its temperature resistance and limit.
  - High temperature will greatly increase the elastomer swelling, decrease its mechanical and chemical properties which will enhance the failure probability of a broken elastomer.
- Its strength and its behaviour in the presence of sand.
  - Good mechanical properties, hardness, tensile strength.
  - Good abrasion resistance.
- Its high chemical resistance and swelling resistance with aromatics and diluents used.
  - Swelling measures the increased elastomer volume as a function of the type of fluids, temperature and time. The chemical swelling is the dissolution of fluids liquid and gas through the elastomer surface, then diffusion in the elastomer matrix. Swelling kinetics is a function of solvent molecule size, free volume, contact surface area, temperature and time.
- Its resistance in the presence of H_2S and CO_2.
- Its resistance with water formation.
- Its blistering resistance under explosive decompression in gassy well conditions.
- Its bond, durable and reliable adherence between the elastomer and the housing.
- Its long-life duration.

When a pump is scheduled to be run into a new field, it is recommended that ageing tests be carried out in a laboratory with elastomer samples.

According to the *in situ* well fluids and conditions, test bench selection of elastomers can predict how elastomers will react to specific well fluid in downhole conditions. Tests enable elastomer dimensional change to be predicted, mainly due to fluid composition, thermal expansion and swelling.

Test benches are essential to fit rotor-stator run life, to optimize PCP pump system efficiency and for reliability.

3.3.3 Elastomer Characterization and Classification

An elastomer stator is physically characterized by its formulation and technology and each PCP's manufacturer supplies its own formulation and specific elastomers properties and expertise.

3.3.3.1 Nitrile (NBR)

Nitrile elastomers (NBR) are copolymers of butadiene and acrylonitrile (ACB), nitrile rubber also referred to as Buna N.
- The acrylonitrile imparts resistance to hydrocarbons, oils and minor gas permeability.
- The butadiene gives properties of sulphur vulcanisation (double bond) and elasticity and flexibility at low temperatures.

Maximum service temperature: 93°C (200°F).

Suitable for heavy to light oils, sand and gas. Limits on API, aromatics, CO_2 and H_2S.

3.3.3.2 Hydrogenated Nitrile (HNBR)

Hydrogenated Nitrile elastomers (HNBR) are copolymers of butadiene-acrylonitrile from which the double bonds coming from butadiene have been more or less totally saturated by selective hydrogenation, without damaging the nitrile functions. They are characterized by:
- An excellent resistance to abrasion.
- Maximum service temperature: 149°C (300°F) water steam.
- An excellent resistance to amine corrosion inhibitors.
- An excellent resistance to acid gas H_2S and CO_2.

Suitable for heavy to light oil, sand and gas, high temperatures and H_2S environments. Limits on API, aromatics and CO_2.

3.3.3.3 Fluorocarbonate (Fluorocarbon) (FKM)

As Viton™ from DuPont Nemours, Fluoro elastomers (FKM) are florine-containing elastomers.

They can operate at higher temperatures than nitrile-based elastomer, characterized by:
- A small swelling in the aliphatic and aromatic hydrocarbons.
- Excellent strength at temperature (permanent elasticity up to 200°C (390°F).
- High temperature applications: maximum 232°C (450°F).
- Poor mechanical performance, lower than nitriles
- Minimal permeability.

- Suitable for light oil, high aromatic content.
- More costly.

3.3.3.4 Alternative Materials Types

Manufacturers may specify other proprietary type specific elastomers as below.

Elastomer Classification

ELASTOMER	MECHANICALS		CHEMICALS			TEMPERATURE Limit Max at pump stator
	Wear & Tear	Abrasion	Aromatics	H_2S	CO_2	
NBR Standard nitrile	Excellent	Good	Medium	Good Medium	Good	120°C/250°F
Soft Nitrile	Good	Excellent	Poor	Medium	Poor	80°C/250°F
HNBR Hydrogenated nitrile Medium Acrylonitrile	Good	Excellent	Medium	Excellent	Excellent	140°C/280°F
High Acrylonitile Saturated nitrile	Good	Medium	Good	Medium	Medium	100°C/210°C
FKM Fluoro carbon Viton type	Medium	Poor	Excellent	Good	Excellent	130°C/270°C

3.3.4 Elastomer Resistance Challenge

3.3.4.1 Temperature Environment

The temperature is the major limitation of elastomeric stators with two sources of heat. The first is the external environment, the well (depth, or specific thermal recovery with steam injection) and the second is from the internal of the pump with friction between the rotor and stator running. Caused by poor inflow and a blocked intake, the pumped fluids do not sufficiently lubricate the rotor-stator resulting in an absence of sufficient cooling and an increase in elastomer temperature.

At high temperatures and over time, most elastomers, including the nitriles typically used in PCP, become hard and brittle and may not form a seal with the rotor. In these conditions, pumps can fail very quickly.

With elastomers currently available on the market. operating temperatures are limited to approximately 150°C (300°F).

This temperature limit will not increase so long as elastomer technology remains a central element of the pump.

3.3.4.2 Chemical Environment

The pumped fluid can contain a high proportion of light aromatics and diluents injected under the pump to reduce the viscosity are currents limitations for PCP.

3.3.5 Elastomer Conclusions

Elastomer technologies are the central element of the PCP stator and a key factor in determining the PCP's run-life.

Standard elastomers currently available on the market are operating temperature limited to approximately 140°C (280°F). This is generally high enough for many wells with PCP downholes.

For higher temperatures and difficult fluids, manufacturer provides specific elastomers.

On-going research into new elastomer formulations, properties and selection will enable higher temperature or problematic light oil wells to be withstood, ensuring an extended lifespan and continued lift operation.

See chapter 17.

3.4 PUMP EFFICIENCY EVALUATION

Bench tests in *"in situ"* in fluid conditions downhole should be carried out, where the pump will be run at rotation speed, differential temperature and pressure.

3.4.1 Calculation of Volumetric Pump Efficiency

The pump efficiency settled in a well is calculated by dividing the fluid volume measured at the separator by the total water volume measured during the bench test under the same pressure, temperature and rotating speed conditions.

Low efficiency is not always the result of a pump wearing out. It may also be caused by a large volume of gas at the pump inlet (see chapter 7).

3.4.2 Evaluation of the Pump's Actual Flow Rate

The pumpable flow rate is determined by a bench test at three different speeds, and by measuring the flow rate, without differential pressure through the pump. The pumped flow rates are then reported to a 100 rpm speed, referring to the standard ISO 15136-1 codification. The average of these three values defines the actual flow rate reference of the pump without differential pressure through the pump.

3.4.3 Internal Recirculation Flow or Slippage Under Pressure

The flow rate drop of under a differential pressure through the pump refers to the volume of leakage inside the pump. This slippage is defined as a reduction of flow due to leakage across the dynamic seal lines between the stator and rotor cavities, compared with the actual pump flow rate reference without differential pressure through the pump.

This slippage is theoretically independent of the rotating speed. It will be measured on the bench by taking the resulting average achieved at the three different rotating speeds at a given head rating.

In the case of viscous oil pumping, the slippage is lower, and the pump efficiency increases with the viscosity of the fluid pumped (70 to 90% production can be reached). So, pumps having with efficiency in the presence of water (low viscosity), can operate satisfactorily with viscous oil.

3.4.4 Pump Efficiency and Fluid Slippage

The PCP, based on the positive displacement principle, is a very simple machine whose performance, efficiency, operating time and life span are determined by some basic parameters. The main parameters which will affect the PCP's performances are:

3.4.4.1 Well bore Conditions

Fluid properties, oil and gas properties, H_2S and high aromatic contents, solid or abrasive concentrations, heavy sand, temperature.

3.4.4.2 Elastomer Selection

Stator elastomer selection is a very important factor for pump efficiency.

The well bore conditions, fluid properties and temperature will determine the correct elastomer compound choice.

The elastomer's mechanical properties and the pump geometry rotor-stator fit can be altered over time by:
- Temperature. Some elastomers may withstand a temperature of 160°C.
- Elastomer stator swell due to aromatic/chemical permeation from gaseous hydrocarbons fluids.
- Elastomer frictional wear due to presence of abrasive products, sand, etc.

Results are higher rotor torque and a loss of pump performance over time.

3.4.4.3 Rotor Rotation Speed

The rotor rotation is limited by rotor imbalance equal to $M\omega^2$ and the friction speed of the rotor in the stator. Progressing Cavity Pumps are driven at low speeds (< 500 rpm) and are therefore not greatly affected by these parameters.

3.4.4.4 Rotor-stator Lubrication

The pumped liquid ensures lubrication between the stator and the rotor. If, for some reason, this lubrication is not achieved, the pump will run dry and will result in elastomer overheating and rapid destruction. The stator is then damaged. To prevent this problem it is recommended to target a minimum 60 meters (200-ft) dynamic fluid above the pump.

3.4.4.5 Rotor Size Selection for a Stator

The selection of the rotor size (rotor O.D.) is an important point to get the best seal and efficiency while running for a specific application to ensure success. The right rotor size will avoid the over-swelling from the start to the end of the pump's life, with expected flow rate and temperature range variation particularly with Cyclic Steam Stimulation.

Manufacturers provide different helical steel rotor size diameters for each pump model to provide a range of rotor-stator combinations and fit for a specific application and downhole conditions in terms of fluid characteristics, pressure, temperature and rate.

For example, there are 10 rotor sizes for each pump model, covering pumping temperatures of 0°C to 155°C.

3.4.4.6 Rotor-stator Fit and Pump Efficiency

The design of PCP's rotor-stator fit is a very important factor in terms of pump efficiency.

From the PCP's principle:

- The rotor is rolling with no sliding inside the stator cavity.
- The steel rotor – elastomeric stator contact line is a compression contact with a positive seal between the metallic rotor and the soft elastomeric stator. The contact line separates two adjacent high-pressure and low-pressure cavities.
- The pump efficiency depends on the leakage between two adjacent cavities. For good pump efficiency, the leakage between two adjacent cavities has to be minimized.
- Rotor wear and corrosion. Rotors are manufactured from specific heat-treated carbon steel chrome plated for corrosion and wear resistance.
- Pressure. Too high pressure and temperature increase the leakage; reduce the pump performance and life span.

Manufacturers design and optimize the rotor-stator profiles and interference fit to maximize pump efficiency, to minimize the drive system's power consumption and to maximize the stator life.

The optimum PCP's efficiency is close to 80% with maximum life span for the better rotor-stator fit and seal. In terms of optimum PCP efficiency, too loose a fit can cause stator wash and too tight a fit can cause high elastomer stress (Fig. 12).

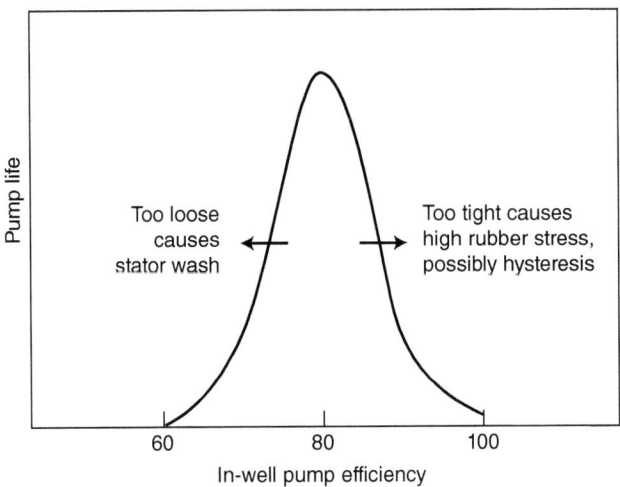

Figure 12

PCP efficiency *versus* rotor fit and pump life.
Source: *OilLiftTechnology*.

CHAPTER 4

Oil Characteristics

4.1 API OIL GRAVITY SCALE

In 1921, the American Petroleum Institute created the API Gravity Scale, initially to measure the specific gravity (SG) of liquids less dense than water, especially petroleum. Use of the API Gravity Scale is now extended to the whole range of specific gravity. The API gravity scale is recognized by the petroleum industry and widely used today. API gravity is graduated in API degrees (API) and the formula used to obtain the API gravity of petroleum liquids is:

$$\text{API gravity} = (141.5/\text{Specific gravity}) - 131.5$$

$$\text{Specific Gravity} = 141.5/(\text{API Gravity} + 131.5)$$

Thus

- Water with a specific gravity of 1 has an API gravity of 10 and 1 cp viscosity.
- The specific gravity and API gravity evolve in opposite directions.
- The heavier or denser the oil, the lower the API gravity is.
- Heavy oil and bitumen are characterized by their high viscosities and high oil specific gravities (SG) or a low API gravity.
- Extra-heavy oil and bitumen with an API of less than 10 means a specific gravity greater than 1 and heavier than pure water.

According to the Canadian Center for Energy, crude oils are classified into different categories according to specific gravity and viscosity at reservoir conditions (Figures 13-1, 13-2):

- **Light crude**, defined as having an API gravity greater than 31.1°API, i.e specific gravity less than 0.87.
- **Medium crude**, defined as having an API gravity in the range between 31.1 and 22.3°API, i.e. specific gravity between 0.87 and 0.92
- **Heavy crude**, defined as having an API gravity in the range between 22.3 and 10°API, i.e. specific gravity between 0.92 and 1.00. In addition to its high specific gravity (equivalent to a low API°grade), heavy crude is defined as having high viscosity, generally above 10 centipoises (cP) and less than 10,000 cP (10 Pa.s) and it flows in reservoir conditions.

- **Extra-heavy oil**, the API degree of which is less than 10 and the *in situ* level of viscosity is less than 10,000 cP (10 Pa.s), which means that it has some mobility in reservoir conditions.
- **Natural bitumen**, often associated with sands, and also referred to as "tar sands or oil sands", the API gravity of which is less than 10 and the *in situ* viscosity greater than 10,000 cP; it does not flow in reservoir condition.

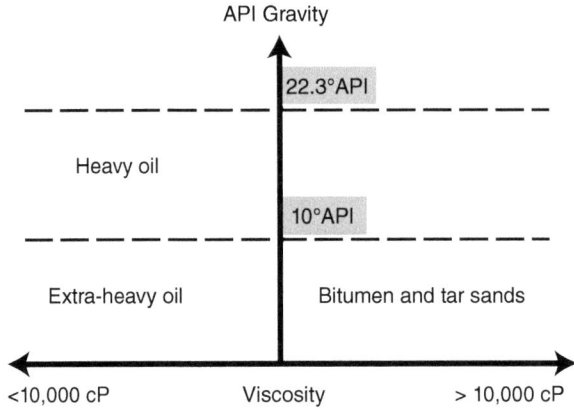

Figure 13-1

Definition of Conventional Heavy Oil, Extra-Heavy Oil and Bitumen.
Source: *American Petroleum Institute & Canadian Center for Energy*.

Oil density (kg/m^3)	Specific gravity	API
780	0.780	48.8
800	0.800	45.3
820	0.820	41.0
840	0.841	36.9
860	0.861	32.9
880	0.881	29.2
900	0.901	25.6
920	0.921	22.2
940	0.941	18.9
960	0.961	15.8
965	0.966	15.0
970	0.971	14.3
975	0.976	13.5
980	0.981	12.8
985	0.986	12.1
990	0.991	11.3
995	0.996	10.6
1000	1.000	10.0
1020	1.012	7.1

Figure 13-2

Specific gravity/API Conversion.

Oil Properties

Most *heavy crude oil* is the result of the bacterial alteration of conventional oil within the reservoir rock. It has different physical and chemical properties, and is generally degraded as compared to conventional light crude: high viscosity and high levels of asphaltenes, heavy metals, sulfur and nitrogen (Figure 13-3) within it. These special properties require specifically adapted technical solutions to produce with artificial lifts, to transport and to refine.

	Zuata	Brent
Gravity (API°)	8.5	38.5
Viscosity (cSt at 60°C)	4,000	4
Sulphur (% by weight)	4.1	0.4
Ni (ppm)	94	1
V (ppm)	450	5
Acidity (mg KOH/g)	4.3	0.05

Figure 13-3

Comparison of the Composition of Extra-Heavy Oil (Zuata Venezuela) and Light Oil (Brent North Sea).
Source: *IFP Energies nouvelles*.

4.2 OIL VISCOSITY AND TEMPERATURE

A measured oil viscosity depends on temperature. Figure 14-1 shows the relationship between viscosity and temperature for oils of various API° gravities. It can be seen that order of magnitude changes in viscosity can be achieved by variations in temperature.

When the viscosity of a crude oil is known at a given temperature (point A), it is possible to obtain an approximate value of viscosity at another temperature by drawing a line parallel to the slanted lines (point B).

The oil viscosity measured in laboratory conditions at a given temperature is not the viscosity of the oil in the formation and the one encountered by the pump in downhole conditions. Live oil contains generally free and dissolved gas, contributing to a much lower real oil viscosity. This apparent viscosity should then be measured in laboratory conditions, in order to evaluate more precisely the frictional pressure drops.

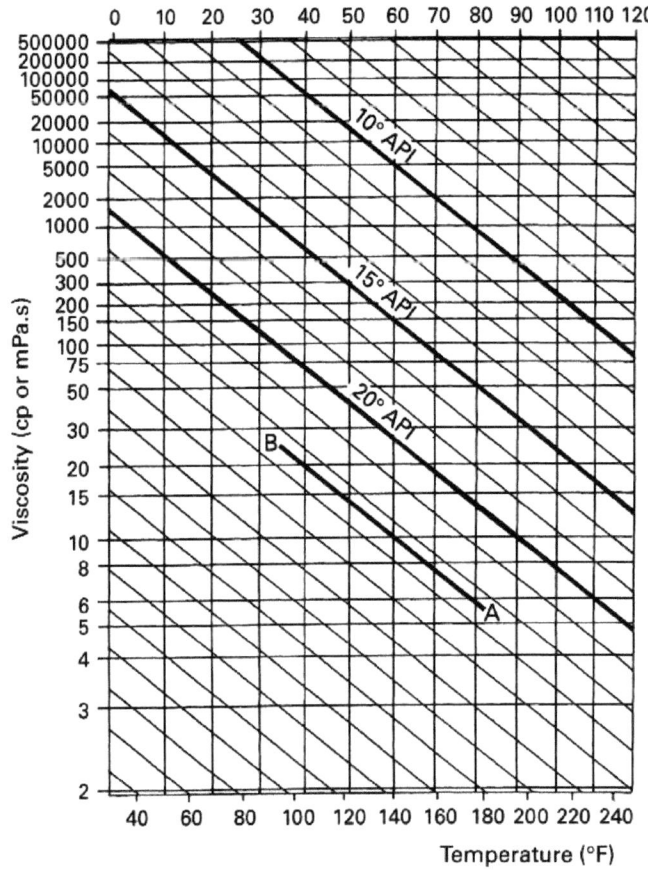

Figure 14-1

Temperature effect on oil viscosity (from Owens and Souler).

4.3 HEAVY OIL AND DILUTION

Generally speaking, the lower the viscosity of the diluent, the lower the viscosity of the diluted oil, however the ratio between the heavy oil and the solvent is important for predicting the viscosity of the blend and simple mixing rules do not apply. The diluent selected should be controlled beforehand in laboratory conditions in terms of efficiency in diluting heavy oils and its compatibilities with asphaltenes to eliminate risks of deposits.

Figure 14-2 shows order of magnitude changes in viscosity of diluted crude by variations of ratio heavy oil/kerosene dilution *versus* temperature.

For example the *80/20 (*read 80% heavy oil 20% kerosene), an injection of 20% kerosene can reduce the oil formation viscosity by 50, as well as for the frictional pressure drops.

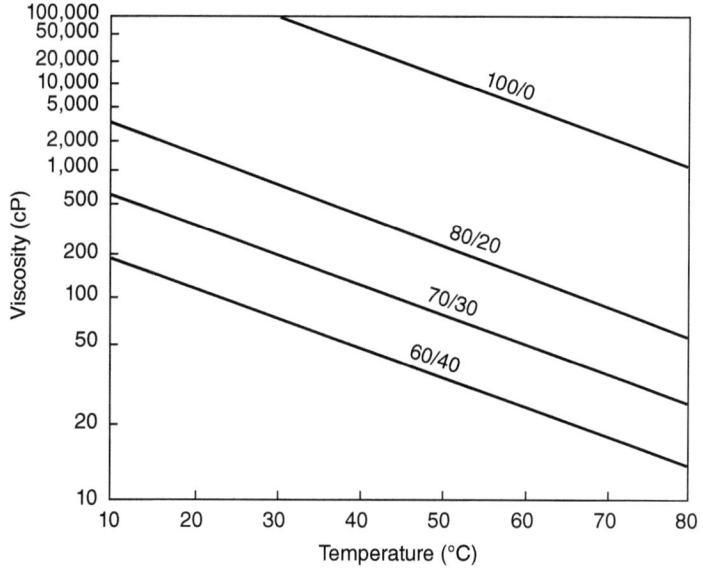

Figure 14-2

Heavy oil/kerosene dilution and viscosity reduction.
Source: *IFP Energies nouvelles*.

4.4 HEAVY OIL AND PCP

It is estimated that 50% of the world's hydrocarbon reserves come from heavy oil fields where viscosity varies from 500 to 15,000 cP. These viscous oils are generally produced from shallow reservoirs. The most significant are located in Canada, Venezuela, Russia and China.

The heavy oil reservoirs operated from vertical wells are not very productive (5 m^3/d). The technique of horizontal drilling and of multilateral wells has resulted in an appreciable increase in well production.

The formation of heavy oil fields is often made up of unconsolidated sand, which gives sand production mixed up with oil and results in beam pumps performing inefficiently.

The introduction of Progressing Cavity Pumps has resulted in this new process for producing heavy oil fields becoming very wide-spread. This acquired experience now means that production can be carried out efficiently in these fields.

To evaluate how easily oil will flow and how it will be lifted by specific PCP, the most important factors are the characteristics of the effluents produced, the well architecture drainage and completion.

CHAPTER 5

Artificial Lift and Selection of a PCP

Artificial lift is used when the pressure in the oil reservoir has failed to the point where a well will not produce at its economical rate using natural energy. This is particularly the case with heavy oil and bitumen reservoirs, because the high viscosity combined with a low reservoir pressure prevents their production by natural pressure drives.

The progressing cavity pump has its own limit, so it is necessary to carefully adapt the pump to a candidate well and operating conditions.

PCP's pumps are chosen according to:

- The reservoir characteristics.
- The characteristics of the effluents produced, CO_2, H_2S, aromatics.
- The well geometry, architecture drainage, horizontal well.
- The well completion and sand control liner screens.
- The PCP sized to well geometry and completion.
- The operating conditions.
  - The frictional pressure drop generated by the viscosity.
  - The resistant torque generated by the viscosity.
- Fluid viscosity reducer.
- Fluid formation gas content.
- Fluid temperature.
- Presence of sand.
- The production criteria.

Each manufacturer offers a range of pumps for various working conditions. For the correct choice, the operator should fill out a data sheet similar to the example shown below in figure 19-3.

5.1 THE RESERVOIR CHARACTERISTICS

The capacity of a well to produce is an essential parameter for its economic implications. Its value then needs to be investigated. However, it is a changing parameter which tends unfortunately to decrease.

The flow rate of a well **Q** depends on:
- The difference between the available pressure which is the reservoir pressure P_g, and the bottom hole back pressure P_f.
- The reservoir and the *in situ* fluids parameters.

For a liquid with a permanent and circular radial discharge, with a fluid speed which would not be too high within the well environment, the Productivity Index **IP** is the ratio of the instantaneous flow rate of the well to the pressure differential between the reservoir pressure P_g and the bottomhole pressure at pump inlet P_f.

The flow rate equation of the well is:

$$Q = IP(P_g - P_f)$$

The productivity index **IP** depends mainly on the fluid viscosity, the formation permeability well surrounding disturbances, and the reservoir height.

Consequently, knowledge of the static level and of the productivity index allows the capacity of the well to be assessed according to an optimum dynamic level.

5.2 THE PRODUCED EFFLUENT CHARACTERISTICS

The petroleum effluents are characterized by their physicochemical properties, and in particular by:
- The viscosity at reservoir pressure and the density (API° or specific gravity) correlated with the viscosity. Generally the more viscous the oil, the heavier it is.
- The free or dissolved gas presence in oil. It is symbolized by the standard GOR gas oil ratio or GLR gas liquid ratio. Gas, in micro bubbles present in the heavy oil, does not form a continuous phase and shows a strong tendency to foam below bubble point.
- The presence of CO_2 and H_2S. Elastomers used to fit stators of oil pumps are resistant to gas, but an initial check is recommended. Elastomers with hydrogenated nitrile function (HNBR) have an excellent hydrocarbon resistance, H_2S and corrosion inhibitors.
- The presence of chemical product aromatics etc. When elastomers come into contact with aromatics, they are more sensitive to swelling. According to the "oil specific gravity/aromatics percentage" pair, the elastomers which are the best suited are selected (elastomer with nitrile function), and the initial interference is adjusted. See chapter 17-1 "Stator elastomer failures".
- The down-hole fluid temperature and pressure at the inlet of the pump and on surface.
- The presence of sand, the formation of heavy oil fields is often made up of unconsolidated sand, which results in a sand production mixed up with oil.

5.3 WELL GEOMETRY, ARCHITECTURE DRAINAGE, HORIZONTAL WELL

In 1978, Elf (now Total) and IFP (now IFP Energies nouvelles) launched a large research project to drill, perform logging and complete four horizontal well onshore and offshore. The offshore horizontal well was at Rospo Mare field, a heavy oil karstic reservoir in the Adriatic Sea. The technical and economical validity of the technology having been proven, in 1988 a boom occurred worldwide onshore and offshore to drill and produce from horizontal wells. Later on improvements in drilling and completion technology led to a horizontal lateral length of more than ten kilometres being drilled and "Advanced wells" with more complex geometries and architectures.

The most common "Advanced wells" are cluster wells (slanted or curved branches drilled with different azimuths from the same vertical hole), stacked wells, multilateral wells (composed of several horizontal arms drilled from the same horizontal drains), re-entry and 3D wells.

The main advantages in producing from horizontal and advanced wells compared to conventional vertical-low deviated wells are the ability to produce either at the same flow rate with lower pressure drawdown, or at greater flow rate with the same pressure drawdown. The results are an increased production rate due to higher productivity and an accelerated oil recovery due to these higher rates resulting in increased oil recovery per well.

PCP artificial lift is part of the success to produce primary heavy oil from complex well architecture, horizontal and advanced wells from many huge heavy oil reserves located in the US (Southern California), Canada (Alberta), Venezuela (Orinoco Belt), Mexico, Columbia, Abu Dhabi, China, Russia, Australia and Africa.

5.4 WELL COMPLETION AND SAND CONTROL LINER SCREENS

Heavy oil field formations are often unconsolidated sand. Sand control liner screens are used to filter out sand and other contaminants from the oil produced, to minimize risk to plug over time and to increase the productive life of the well.

Sand control screen completions are many used to produce heavy-oil reservoirs on primary cold recovery and on thermal recovery.

On cold production, slotted liners are the most common used at the lowest cost.

On thermal recovery the heavy oil, heated to allow it to flow with lower viscosity toward the wellbore, brings large volumes of sand to the wellbore interface naturally. The sand control screen liners used are slotted, metal-mesh, wire-wrap and pre-pack gravel pack.

Tubular slotted screen liners from 2 to 16 inches are machined with multi-high-speed steel circular saw cutting blades to make straight or keystone slots. Straight slots are the same width on the outside and the inside of the pipe. Keystone slots are wider on the inside than they are on the outside. They are self-cleaning, larger grains are stopped on the outside, smaller grains are produced. Slot geometry and density depends on sand grain sizes, heavy oil characteristics and production rate.

Sand control screens preserve the PCP's run life and efficiency, increase well production with lowered operating costs per barrel.

5.5 PCP SIZED TO THE WELL GEOMETRY AND COMPLETION

In accordance with well geometry, well completion casing and tubing sizes, pump flow rate and pressure, PCP outside diameters ranging from 1-1/2" (1.4 m³/day at 100 rpm) to 7" (300 m³/day at 100 rpm) are used in respect of rules.

The **stator** must:

1. Be able to go through the casing and any other part which can be integrated into the casing equipment. Especially in very deviated wells, it must be ensured that the pump is perfectly fitted into the curved casing profile.

2. If necessary, reserve enough annular space with the casing, in order to run down servicing or fishing tools eventually, and the setting of a gas separator.

3. Enable preservation of a sufficient annular space, if the pump is installed below the perforations.

The **rotor** must be able to go through the tubing or any integrated part.

A sufficient diameter is necessary in the tubing, so that the rotor's eccentric motion can turn around inside the bottom part of the drive strings.

5.6 PCP'S OPERATING CONDITIONS

Two main parameters to operate PCP's downhole pumps are the frictional pressure drop and the resistance torque on the drive string.

5.6.1 Frictional Pressure Drop Generated by Viscosity

Heavy and viscous oils lead to frictional pressure drops in the tubing; hence a decrease in the performance of the pump as far as the fluid height to be lifted is concerned. So, these pressure drops have to be evaluated and a higher head rating pump should be chosen.

The discharge of high viscosity fluids through the production tubing can generate significant pressure drops which are proportional to the oil viscosity.

$$\Delta P_f = \frac{7.05 \times 10^{-4}}{(D+d)(D-d)^3} \times Q \times \mu_f \times \frac{1}{\ln \frac{\mu_s}{\mu_f}} \left(\frac{\mu_s}{\mu_f} - 1 \right)$$

ΔP_f pressure drop due to friction (in bar)
D inside diameter of the tubing (in cm)
d drive string diameter (in cm)
Q pumped flow rate (en m³/day)
μ_f viscosity of the effluent at the inlet temperature (in cP)
μ_s viscosity of the effluent at the surface (in cP)
L length of the tubing (in meters)

5.6.2 Resistant Torque Generated by the Viscosity

Also, the viscosity increases the value of the resistant torque of the PCP's drive strings in the tubing.

$$\Gamma_v = 0.165 \times 10^{-8} \times \mu_f \times L \times N \times \frac{d^3}{D-d} \times \frac{1}{\ln \frac{\mu_s}{\mu_f}} \left(\frac{\mu_s}{\mu_f} - 1 \right) \quad \text{where:}$$

Γ_v resistant torque (in daN.m)
μ_f viscosity of the effluent at the inlet temperature (in cP)
μ_s viscosity of the effluent at the surface (in cP)
L length of the tubing (in m)
N rotary speed (in rpm)
d drive string diameter (in cm)
D inside diameter of the tubing (in cm)

These two relations, pressure drop and resistance torque due to friction in tubing, point out the influence of viscosity and temperature. As a matter of fact, if the formation oil is at the bottomhole temperature, at the pump inlet, this temperature will decrease in accordance with the geothermal gradient. The viscosity increases perceptibly, as the pressure drops along the tubing to surface. Figure 14-1, which represents the viscosity variation according to the temperature, shows that a variation of temperature from 60 to 40°C increases the viscosity from 5,000 cP to about 40,000 cP, or 8 times. A pressure variation generated by an increase or a decrease of pressure drops will be expressed by output power variations of the drive motor.

The evaluation of the pressure drops and of the resistant torque due to the viscosity is then very important for selecting:

– The pump (admissible head rating).
– Drive strings (admissible torque).
– The motor (required power).

5.7 FLUID VISCOSITY REDUCER

Fluid viscosity has an impact on PCP's efficiency and there are several solutions for reducing pressure drops generated by the oil viscosity. When it is possible:

1. **Use tubing of a larger diameter**. The above relations show that pressure drops due to friction vary to the power of 4 of the internal tubing diameter D. Nevertheless, the tubing diameter is limited by the internal casing diameter, and a flow speed which will have to take into account the sand sedimentation possibly transported. However, when the oil is very viscous, the sedimentation speed of the sand is low.

2. **Use high capacity pumps** running at low rotating speeds.

3. **Use continuous drive strings**, without sub-coupling which creates restrictions.

4. **Insulate tubing** near the wellhead in cold areas, or during the cold seasons of the year.

5. **Reduce the downhole viscosity** fluids by injecting diluent.

The blending of heavy oil with light oil, gas condensate, refinery naphtha or kerosene, water with additives) reduces the downhole oil viscosity. See chapter 4-3.

The diluent injection should be made if possible at the pump inlet in order to help the mixing of the various oil specific gravities. The elastomer behaviour with the oil/diluent mixture should be controlled before in the laboratory. See chapter 3-3.

To conclude on the PCP's selection for a well candidate and an expected production rate, it is better:

- To use a wider tubing diameter with a high capacity pump running at a low rotating speed.
- To minimize all restrictions in tubing.
- To reduce the downhole oil viscosity by using diluent injected at the pump inlet.

In order to increase the life span of the pump, to reduce pump failure and a capital cost saving.

5.8 FLUID FORMATION AND GAS CONTENT

This gas content is symbolized in petroleum production by:
- GOR: Gas Oil Ratio.
- GLR: Gas Liquid Ratio.

Following its principle, the PCP can absorb multiphase effluents (oil, water, gas) with high GOR at inlet, and should be installed in the well above the bubble point. In such a case, the pumped volume (oil + gas) will be higher than the oil volume produced at the surface.

Figure 15 is a representation of the evolution of the pumping conditions according to the GOR, and the influence of the pump position according to the dynamic level and bubble point. If it is desirable to keep the same oil flow rate whatever the pump level, it is necessary

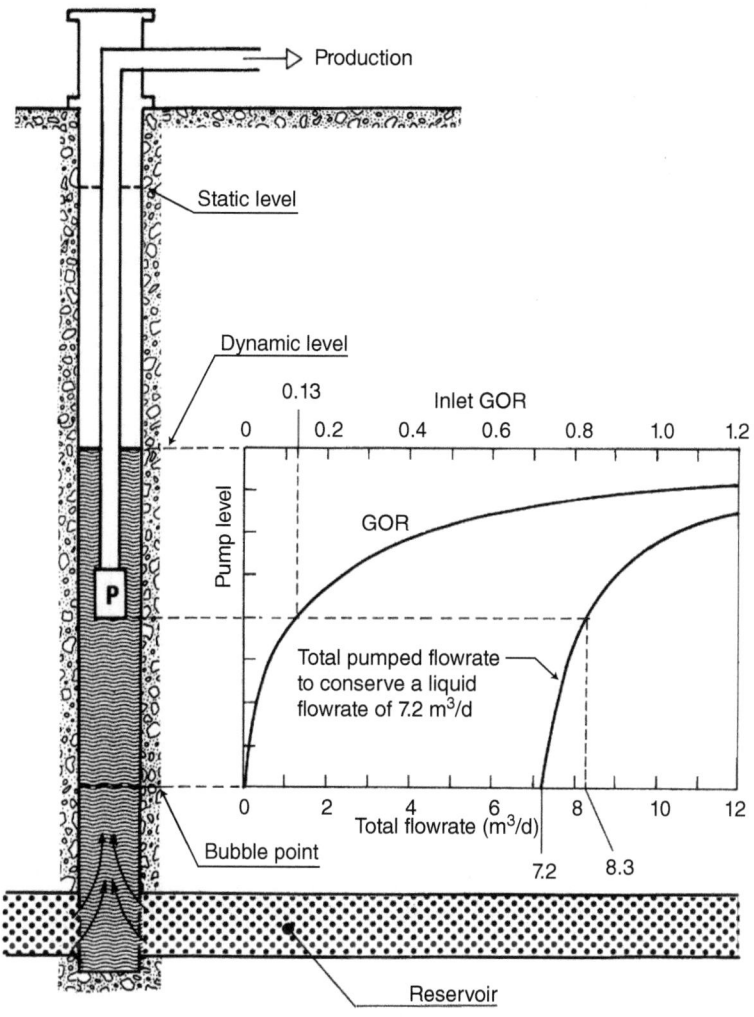

Figure 15

Incidence on the flow rate of the PCP position in an oil well containing gas.
Source: *IFP Energies nouvelles.*

to increase the rotating speed, according to the GOR. For example, Figure 15 reports that at GOR = 0, the oil flow rate is about 7.2 m³/d. But, if the pump is positioned at a higher level, GOR = 0.13 (as indicated on figure), it is then necessary to run it at a higher speed in order to generate a flow rate: oil + gas = 8.3 m³/d, to conserve the same oil production of 7.2 m³/d.

Chapter 7 refers more particularly to the presence of gas at the pump inlet and presents solutions of gas separation. A calculation method allows the evaluation of GOR at the pump inlet.

5.9 FLUID TEMPERATURE

Generally, the Progressing Cavity Pumps whose well installation does not exceed 2200 m, run at temperatures below 120°C (with nitrile stator elastomer). However, some elastomers are able to withstand temperatures of 150°C. The hydrogenated nitriles have good continuing heat strength. Their thermal strength increases with the hydrogenation rate.

The temperature evolution generates large variations of heavy oil viscosity, as shown in Figure 14-1. The oil formation viscosity is important data for the evaluation of frictional pressure drops and the resistant torque of the drive strings.

5.10 FLUID AND PRESENCE OF SAND

5.10.1 Sand Transportation in Production

In the production of heavy oil fields enclosed into unconsolidated sand formations, the prevention of sand inflow cannot be made without changing the production ratio. During the start-up of a production well, a 30% sand ratio can be noted, but later on, this sand ratio is often stabilized at less than 3%.

The presence of sand results in equipment wearing out quicker, in an increase in the drive torque and of engine power. It can also generate sand accumulations at the pump inlet which may necessitate lifting the downhole equipment.

Due to their design, the PCP can withstand moderate sand ratios. Problems are mainly generated by high variations in the rotating speed involving a dynamic pressure decrease and consequently of the pressure caused by the apparition of "sand packets". In order to get a better unrestrained sand flow, it is recommended to vary the pump rotating speed slowly (by steps of 20 rpm and per day).

Sand accumulation above the pump and inside the tubing is a usual problem. It happens after production stops or because of an insufficient pumped fluid velocity, as far as the sedimentation speed of the sand is concerned. The viscous fluids slow down the sedimentation speed of the sand, but generate a frictional increase.

The operator often sets the pump under the casing perforation in order to enable a natural separation of the gas. But, on the other hand, this situation can generate sand accumulation at the pump inlet.

5.10.2 Incidence on the Pump

The abrasion strength of the elastomers is characterized by their resilience. In general, the abrasive resistance is inversely proportional to the hardness of the elastomer. In contact with a particle, the elastomer deforms and takes back its initial shape with little or no wearing.

This unique characteristic of elastomers gives them superiority over metallic parts for abrasion resistance; see chapter 17 Pump Failures.

Progressing Cavity Pumps can produce fluids containing sand. However, a certain number of precautions must be taken:

- Avoid sand packets at the pump inlet, in protecting the pump inlet orifices.
- Ensure, during pumping stops, that there is no sand sedimentation in the tubing (restarting may be difficult).
- If any sand is present, the pumps are designed with rotors having a thick chromium coating and a very elastic elastomer.

5.10.3 Application Limits

Figure 16 defines the application limits of pump models when there is sand present.

They refer to the rotating speed reduction and the admissible head rating. Consequently, the operator must choose higher characteristic models.

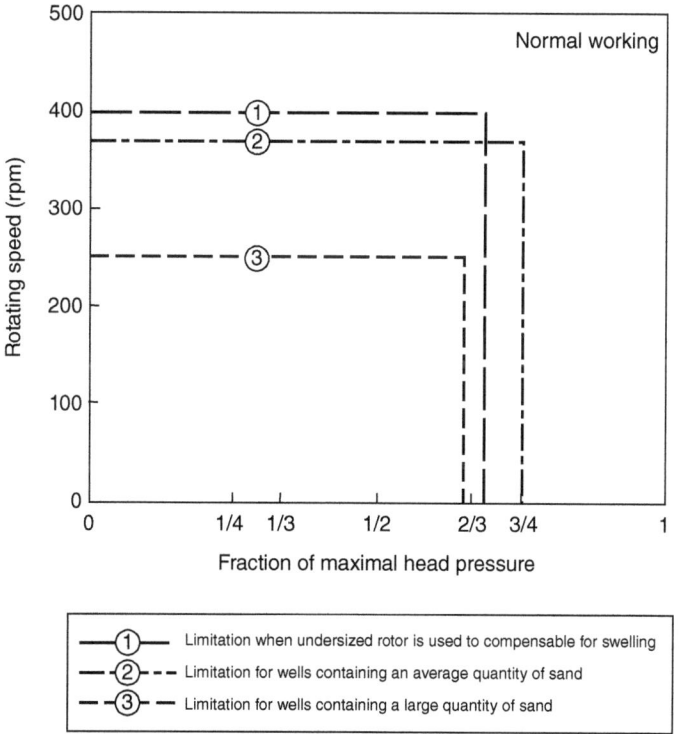

Figure 16

Recommended limitations of working for wells containing a quantity of sand.
Source: *PCM*.

5.11 PRODUCTION CRITERIA

The selection of the pump depends on its position in the well:
- The dynamic level or submergence level.
- The bubble point level.
- Evaluation of the minimum head rating of the pump.
- The admissible head rating of the pump (column height to be discharged, pressure drops due to friction generated by the effluent viscosity, and
- The wellhead pressure.

5.11.1 Positioning Level According to the Dynamic Level (or Submergence Level)

In spite of its self-priming characteristics, a sufficient submergence (about 100 m) has to be maintained above the pump, because if dry rotating were accidentally to happen, the pump's stator would be damaged.

It is therefore necessary to properly evaluate the level of submergence of the pump (Sonolog measurement or bottomhole pressure sensor) and to adapt the rotating speed of the pump to the flow rate, allowing the predetermined level to be maintained. Figure 17 illustrates the influence of the rotating speed on the submergence level.

The relation defining the productivity index shows that the dynamic level variation is proportional to the pump flow rate and consequently to the rotating speed. However, when the pump level is above the bubble point, the presence of free gas is not preserved in the same proportions.

The best case is the possibility to run down the pump below the reservoir bed, so any danger of dry rotating would be excluded, and a natural separation of gas would occur, which is good efficiency of the pump.

5.11.2 Positioning Level in Respect of the Bubble Point Level

In order to generate maximum efficiency, it is always preferable to place the pump below the bubble point level. However, in certain circumstances (limitation of the pressure drops in the tubing, high temperature, well profile…) it may be decided to place the pump above this critical level.

Figure 15 is an example which makes it possible to know, according to the pump level, the GOR and the total flow rate at the inlet compatible with the flow rate required at the surface. Described in paragraph 4.4.1, the calculations demonstrate on this figure the necessity to run at a higher speed when the pump is positioned above the bubble point thus to keep the same flow rate.

5.11.3 Evaluation of the Pump's Minimum Head Rating

It is the sum of:
- The pressure generated by the column height to be discharged (between the surface and the submergence level).
- The pressure drop due to friction along the tubing (high if the pumped fluid is viscous).
- The pressure required at the wellhead.

The selected pump type must therefore be able to generate a head rating higher than the sum of the three values listed above.

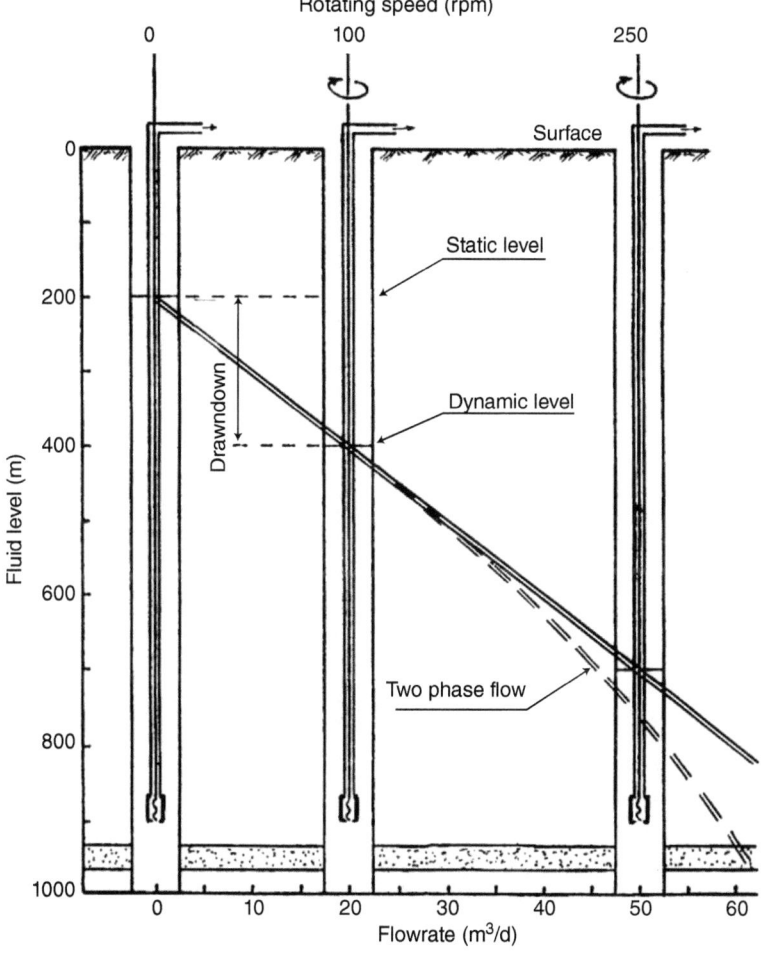

Figure 17

Relationship between the rotating speed and the fluid level.
Source: *IFP Energies nouvelles*.

5.11.4 Pressure Generated by the Column Height to be Discharged

This pressure depends on the column height to be discharged and on the specific gravity of the effluent.

N.B.: the specific gravity of the fluids may be different in the annulus and inside the tubing (influence of the gas included in the oil).

5.11.5 Pressure Drop Generated by the Viscosity of the Effluent

They are evaluated according to the relationships that are commonly used for drilling fluids and by taking into account an increase of viscosity between the bottomhole and the surface, due to a decrease in the ambient temperature along the tubing.

$$\Delta P_f = \frac{7.05 \times 10^{-4}}{(D+d)(D-d)^3} \times Q \times \mu_f \times \frac{1}{\ln \frac{\mu_s}{\mu_f}} \left(\frac{\mu_s}{\mu_f} - 1 \right)$$

ΔP_f pressure drop due to friction (in bar)
D inside diameter of the tubing (in cm)
d rod string diameter (in cm)
Q pumped flow rate (in m³/day)
μ_f viscosity of the effluent at the inlet temperature (in cP)
μ_s viscosity of the effluent at the surface (in cP)
L length of the tubing (in m)

As shown Chapter 4, Heavy Oil and Dilution, Figure 14-1 relates the variation of the viscosity *versus* temperature and on Figure 14-2 the influence of the kerosene dilution on the viscosity. For example, a 20°C increase in temperature in the well reduces the viscosity by a factor of 5 and consequently, the frictional pressure drops.

Proper knowledge of the characteristics of the pumped fluid is therefore very important, especially since the presence of gas in oil influences the viscosity value.

5.11.6 Wellhead Pressure

The wellhead pressure depends on the characteristics of the surface system:
- Length and diameter of the pipelines.
- Viscosity of the transported fluid.
- Flow rate.

5.11.7 Head of Pressure Related to Down Hole Artificial Lift

Manufactures' identifications of PCP models (chapter 6-4.1) express the displacement volume in m^3/day or bbl/day related to the rotor speed 500 or 100 or 1 rpm and a head capability express in metre elevation water column and zero head pressure (0 bar). Example from PCM, the PCP model 1000TP600 is rated 1,000 m^3/day at 500 rpm and 600 m (1,975 ft) head pump lift capability.

The "Pressure Head" or "Head" is the term used instead of Pressure because:
- The pump will pump any liquid to a given height or head.
- The amount of pressure depends upon the weight (specific gravity SG) of the liquid.
- Pump manufacturer does not know what liquid the pump will be pumping so he gives only the Maximum Pump Head capability (MPH) or "Head" that the pump will generate expressed in metre (or ft) of water column (mwc or ftwc).
- The Operating pump head recommended = 2/3 of Maximum Pump Head capacity in real operations conditions.
- The **Head** (in mwc or ftwc) is always related to the true vertical depth (TVD) of the well.
- The **Net Lift (NL)** (in metre or feet) = the true vertical depth of the pumping fluid dynamic level = True Vertical depth of the pump, *minus* the True Vertical Depth height of the dynamic fluid level above the pump.
- The **True Vertical head water column of a fluid** (mwc or ftwc) = The Net lift (True Vertical depth pumping fluid level) × fluid specific gravity (SG).
- The **Head loss** (mwc or ftwc) is the pressure head drop due to friction loss of fluid up through tubing, piping, valves and fitting expressed in water column pressure head.
- The **Flow line back head pressure** (mwc or ftwc) is the pressure head required at the surface well head or the flow line back head pressure.
- The **Total Dynamic Head of a fluid (THD)** or **Total Net Lift (TNL)** in mwc or ftwc.
  - **Total Net Lift** = the total head required when pumping at the desired rate. It is the difference between the head at the pump discharge and the head at the pump intake assuming the produced fluid has no free gas, and all the pressures are converted to head using the specific gravity of the produced fluid.
  - **Total Net Lift** = the Net Lift (at pumping fluid level) × fluid specific gravity (SG) + Flow line back head pressure (mwc or ftwc) + Head Loss or friction loss of fluid up the tubing (mwc or ftwc).

Pressure at the suction of the pump or pump inlet:
- The Net Positive Suction Head (**NPSH**) is a factor designed into the pump and measurable in test laboratory by the manufacturer. NPSH is the amount of external pressure available to the pump inlet as long as the pump sees continuous-phase liquid.
- The **NPSHr** (required) is the amount of external pressure required to insure the pump operate full liquid. NPSHr is very dependent upon the liquid properties at downhole conditions, mainly the boiling point, gas solubility, vapour pressure, temperature.
- The **NPSHa** (available) is the term for providing sufficient pressure at the pump suction or inlet to prevent "boiling". It is a function of the pumping system and consists

of pressure of the fluid at the pump inlet, the losses in the suction piping and pump inlet and vapour pressure of the liquid at downhole conditions. NPSH needs to be grater than NPSHa.

The PCP rotary positive displacement pump needs low NPSH compared to others down hole pumps such as ESP centrifugal pump, jet pump, rod pump.

5.11.8 Head of Pressure of Water and Fluid formulas

Conversion:
- 1 metre = 3.281 ft, 1 ft = 0.304 m
- 1 bar = 100 kPa = 14.5 psi, 1 kPa = 0.145 psi
- 1 psi = 6.89 kPa = 0.0689 bar
- Fresh water specific gravity = 1 at 4°C (39°F)
- Fresh water gradient = 9.8 kPa/metre = 1.421 psi/m = 0.433 psi/ft
- From Water Pressure to Head Water column in metre (mwc)

$$\text{Head Pressure (mwc)} = \frac{\text{Pressure (kPa)}}{9.8} = \text{Pressure (kPa)} \times 0.102 = \text{Pressure (bar)} \times 10.2$$

$$\text{Head Pressure (mwc)} = \frac{\text{Pressure (psi)}}{1.42} = \text{Pressure (psi)} \times 0.704$$

- One metre of water column (mwc) = 3.281 ftwc = 9.8 kPa = 1.42 psi
- 10.2 metres of water column (mwc) = 33.46 ftwc = 100 kPa = 1 bar = 14.5 psi

From Water Pressure to Head Water column in foot (ftwc)

$$\text{Head Pressure (ftwc)} = \frac{\text{Pressure (kPa)}}{2.98} = \text{Pressure (kPa)} \times 0.3347 = \text{Pressure (bar)} \times 33.47$$

$$\text{Head Pressure (ftwc)} = \frac{\text{Pressure (psi)}}{0.432} = \text{Pressure (psi)} \times 2.32$$

- One foot of water column (ftwc) = 0.304 mwc = 0.432 psi = 2.98 kPa = 0.0298 bar

From Fluid Pressure to Head Water column in metre (mwc)

$$\text{Fluid Specific gravity} = SG$$

$$\text{Head Pressure (mwc)} = \text{Pressure (kPa)} \times 0.102 \times SG = \text{Pressure (bar)} \times 10.2 \times SG$$

$$\text{Head Pressure (mwc)} = \text{Pressure (psi)} \times 0.704 \times SG$$

From Fluid Pressure to Head Water column in foot (ftwc)

$$\text{Head Pressure (ftwc)} = \text{Pressure (kPa)} \times 0.3347 \times SG = \text{Pressure (bar)} \times 33.47 \times SG$$

$$\text{Head Pressure (ftwc)} = \text{Pressure (psi)} \times 2.32 \times SG$$

CHAPTER 6

Operational Conditions

The pump model is chosen according to its:
- Flow rate and the head rating.
- Rotor rotating speed.
- Performance and controls.
- Identification, diameter and the length of the equipment.
- Drive systems.

Bench tests may determine the pump performance in conditions close to the wells. However, depending on the results, additional tests may be carried out with rotors of different diameters.

In unusual conditions of use, samples of elastomers should be tested in laboratory conditions in order to evaluate the swelling and other physicochemical characteristics.

6.1 FLOW RATE AND HEAD RATING

The most important criteria are the flow rate and the total head rating. Each manufacturer asks their customers to complete a data sheet form (see PCP data sheet Figure 19-3) which will enable the appropriate pump to be selected for the candidate well.

It is necessary to consider a certain loss in production, generated by the fit quality of the rotor/stator. Down hole temperature and the volume of free gas at the pump inlet influence the rotor/stator interference and contribute to a more or less important leak rate.

The head rating required for a pump may be approximately determined on this basis:
- The dynamic level of fluid in the well.
- The positioning level of the pump.
- The specific gravity of the pumped fluid.
- The frictional pressure drop along the tubing.
- The wellhead pressure.

6.2 ROTOR ROTATING SPEED

The optimum speed of the pump depends on its verticality or its gradient, of its position in the well vertical, slanted to horizontal, as well as the viscosity of the pumped fluid. In every case, it is better to choose a slow rotating speed in order to increase the life span of the pump.

The pump flow rate is proportional to the rotating speed. However, the initial leak rate has to be taken into account, which depends on the head rating required from the pump. Consequently, there is a minimum rotating speed before the pump starts the flow.

Figure 18 represents the variation of the flow rate and the torque depending on the rotating speed for a type of pump at a different head rating.

In the particular case of heavy oil production, operation at the lowest possible speed is recommended in order to increase the life span of the pump, the drive strings, the tubing and the surface equipment. Frictional pressure drops and the specific gravity increase due to the presence of sand must be taken fully into account.

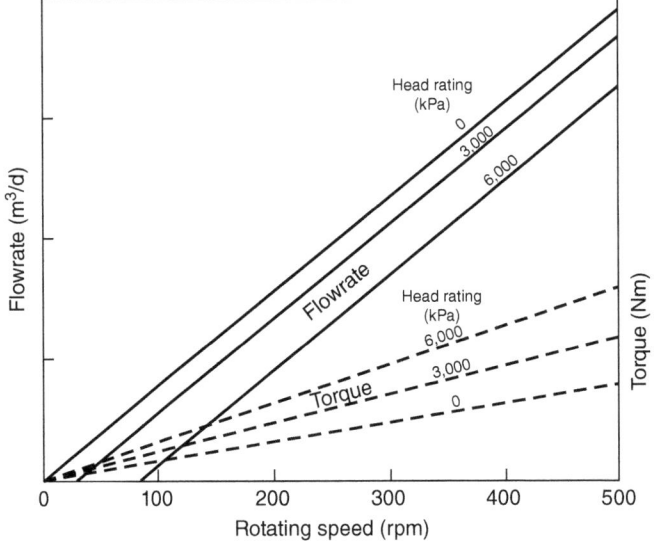

Figure 18

PCP performance curves.
Source: *ISO15136-1*.

6.3 PERFORMANCE CONTROLS AND TESTS

Every pump should be tested before installation. The purpose of the test is to ensure the pump's operational performance.

6.3.1 Test Procedure

The pump is installed horizontally on a bench. Rotation and power are provided by an electric motor. Tests shall be carried out with pumped water through a closed loop system. A choke is used to regulate the discharge pressure which creates a differential pressure across the pump. The volumetric efficiency of the pump is therefore determined.

The ISO 15136-1 standard defines the test procedure, the resulting precision and the acceptance criteria.

The operating method may vary depending on the manufacturers or the users. The following procedure is suggested:
- Vary the head rating by operating at constant speed from zero up to the maximum pressure of the possible use by intervals of 3000 kPa (30 bar). This test procedure should be repeated at several different speed levels in order to generate a sufficient number of points allowing graphs to be drawn according to the model shown in Figure 18.
- The tests are carried out at the same temperature as that of the wells, at the scheduled level of the pump installation.
- To try a pump for a better characterization in working conditions, tests may be carried out with fluids than other water (reservoir oil). But the water test remains the only reference.

6.3.2 Test Report

The test reports for a new pump will state at least:
- The rotating speeds used for tests.
- The measured flow rates corresponding to the fixed head ratings.
- The resulting volumetric efficiencies.
- The test temperature.

6.4 PCP GUIDELINE AND IDENTIFICATION

Progressing Cavity Pumps rotated by drive strings from the surface are identified in accordance with the established ISO 15136-1: 2001(E) standard, but pump manufacturers also use proprietary pump identifications and specifications

There are several steps involved in properly selecting a Progressing Cavity Pump.

6.4.1 Pump Terms and Selection

Pump Terms

1. Pump displacement volume rate at 500 rpm rotor and zero head pressure (0 bar), rate expressed in cubic meters per day (m^3/day) or barrel per day bbl/d (1 m^3 = 6.28 bbl, 1 bbl = 0.1589 m^3, 1 bbl/d = 0.0662 m^3/h).

2. Pump pressure rating or Maximum Pump Head capability (MPH) is expressed in meters of water column (meters of H_2O, mwc) or feet of H_2O.

3. Operating pump head = 2/3 of Maximum Pump Head capacity in real operations conditions.

4. Flushing pump, the outside diameter O.D in mm or inches.

5. Stator pump length in meters or feet.

6. Rotor pump length in meters or feet.

7. Stator pump connection, pin or box API thread type and size.

The well characteristics will determine the pump sizing and selection with proper materials.

6.4.2 PCP Rotor and Stator Identifications

The rotor and stator pump models from ISO 15136-1: 2001(E), are identified with a special nomenclature, the first part refers to an approximation to the flow to be pumped in m^3/day at **500 RPM at zero elevation**, the second part refers to the nominal lift in meters.

Notice: Manufacturers' identifications of pumps rotor and stator are also expressed in **m^3/d at 100 rpm**, or in **m^3/d/rpm**.

6.4.2.1 ISO Stator Code

The stator shall be permanently impressed with the following code:

$$vvv/hh/L_r/cc/eee$$

where:

- **vvv** is the displacement (in m^3/d at 500 rpm rotor speed.).
- **hh** is the maximum head rating of the pump (in MPa).
- **L_r** is the number of lobes of the rotor.
- **cc** is the number of cavities.
- **eee** is the manufacturer's code for the elastomer.

6.4.2.2 ISO Rotor Code

According to the standard ISO 15136-1: 2001(E), the rotor head is identified with the following code:

$$vvv/hh/yyy$$

where:

- **vvv** is the displacement (in m^3/d at 500 rpm rotor speed.).
- **hh** is the maximum head rating of the pump (in MPa).
- **yy** is the length of the rotor protruding below the bottom of the elastomer when the rotor head is at the top of the elastomer.

For example, the PCM model 300TP1800:

vvv = 300 is the capacity: 300 m^3/day at 500 rpm rotor speed, at zero **head pressure (0 bar), or zero** elevation.

hh = 1800 is the nominal lift: 1800 meters of water column (mwc).

6.4.3 PCP Manufacturers' Model Identification

PCP pump model, nomenclatures or identifications vary from manufacturer to manufacturer, for example:

From PCM: **AAA-BB-CCC (Metric)**
- AAA = flow in m^3/day at 0 bar and 500 rpm rotor speed (TM past code) and 100 rpm rotor speed (E new code).
- BB = PCM Moineau™ is TP or MET (metal stator) past code and E new code.
- BB = K for KUDU.
- CCC = the maximum Head Pressure Capacity in meters of water column (mwc).

Example:

The ISO 300-TP-1800:
- 300 = 300 m^3/day, the flow in m^3/day at 0 bar and 500 rpm rotor speed.
- TP is the PCM Moineau™ past code.
- 1800 = 1800 mwc, the max Head Pressure Capacity = 180 bar.

The NEW PCM pump code is 60-E-1800, equivalent to the ISO 300-TP-1800:
- 60 = 60 m^3/day, the flow in m^3/day at 0 bar and 100 rpm rotor speed.
- E is the PCM Moineau™ new code.
- 1800 = 1800 mwc = 180 bar, the max Head Pressure Capacity.

With PCM Vulcain™ with metallic stator: 550 MET 750:
- 550 = 550 m^3/day, the flow in m^3/day at 0 bar and **500** rpm rotor speed.
- MET is METallic stator PCM Moineau™ past code.
- 750 = 750 mwc = 75 bar, the max Head Pressure Capacity.

From PROTEX: AA-BB-CC (Metric)

- AA = maximum head pressure measured per 100 meters of water column = × 10 bar.
- BB = stator tube OD in the PCP model.

Letters marking Stator OD tube	Outside diameter of the stator	
	inch	mm
PC	3.13	79.50
PT	3.50	88.90
PH	3.75	95.25
PX	4.50	114.30

- CC = production rate per day at 100 rpm rotor speed.

Example 12-PX-31: 1200 m maximum head = 120 bar, OD = 4.5", 31 m^3/100 rpm.

From NETZSCH NEMO™ PCP: AAA-BBB-CCC-D_1D_2-E (Metric)

- AAA = NTZ NETZSCH tubular stator or NTU uniform wall stator.
- BBB = Stator diameter (inch) example 500 = 5.00".
- CCC = max Head Pressure Capacity in kgf/cm^2 or bar.
- D_1 = S for single lobe geometry 1-2 or D for Multi-lobe 2/3.
- D_2 = T-tubular stator, S-submerged, HS-hydraulic submerged, IT-insertable, DS-direct drive submerged, TM-metal stator, TS-segmental stator.
- E = displacement (m^3/day) at 100 rpm rotor speed and zero differential pressure.

Example, the NTZ 350 180 ST 25:

- NTZ is a tubular stator.
- 350 = 3.5" stator OD.
- 180 = 180 bar = 1800 meters, the max Head Pressure Capacity.
- ST for single lobe geometry 1-2.
- 25 = 25 m^3/day at 0 bar and 100 rpm.

From Dyna-Lift   DD-AA-CC (Metric)

Example: DD-98-1600

DD = DynaLift, 98 m^3/day at 0 bar and 100 rpm and 1600 m max Head Pressure Capacity.

From Baker-Lifteq Pumps   AA-B-CC (Metric)

AA = flow in m^3/day at 0 bar and 100 rpm rotor speed

B = code type B, D, F, G

CC = pressure in bar = × 10 meters

Example: 223-G-90 = 223 m^3/day at 0 bar and 100 rpm, code G, 90 bar = 900 m max Head Pressure Capacity.

From EUROPUMP   AA-E-CC (Metric)

AA = flow in m³/day at 0 bar and 100 rpm rotor speed

E = Europump code

CC: CC × 10 = max pressure in bar or CC × 100 = xx meters max head lift capacity

Example: 120-E-11.5 = 120 m³/day at 0 bar and 100 rpm, code E,

11.5 = 115 bar max pressure or 11.5 × 100 = 1150 m max head pressure capacity.

From NOV Monoflo   AA-CC (Metric)

AA = flow in m³/day at 0 bar and 100 rpm rotor speed

CC = max head lift capacity in meters

From OILLIFT   AA-CC (Metric)

AA = flow in m³/day at 0 bar and 100 rpm rotor speed

CC = max head lift capacity in meters

From R & M   AA-BB-CC (SL) (Metric)

AA is AA × 100 = max head lift capacity in meters

BB = code pump

CC = flow in m³/day at 0 bar and 100 rpm rotor speed

From CAMERON   CAM-BB-CC-D (Metric)

CAM = CAM-PCP™

BB = flow in m³/day at 0 bar and 100 rpm rotor speed

CC = max head lift capacity in meters

D = code pump

Remarks:
- In the previous ISO identification PCP the flow rate was m³/day at 0 bar and 500 rpm rotor speed. The new flow rate identification is at 100 rpm rotor speed.
- Identification pumps change from manufactures.
- The pressure in meters of water = 1 mwc = 0.1 kgf/cm^2 = 0.1 bar.

6.5   PCP DATA SHEET FROM MANUFACTURERS

Pump manufacturers provide pump specification sheets and characteristics.

A PCP's example from Oil Lift:

Model 200 (1250) – 900 and 1200

Pump flow rate 200 m³/day (1,250 bld) at 100 rpm

Pump lift 900 m and 1200 m

Graphs volume efficiency (%) Per cent of rated pump lift and Torque *versus* rpm rotor shown in Figure 19-1.

Figure 19-1

Example of pump manufacturer specification sheet.
Model 200 m³/day at 100 rpm or 1000 m³/day at 500 rpm, 300 m and 600 m pump lift.
Source: *OilLift Technologies catalogue*.

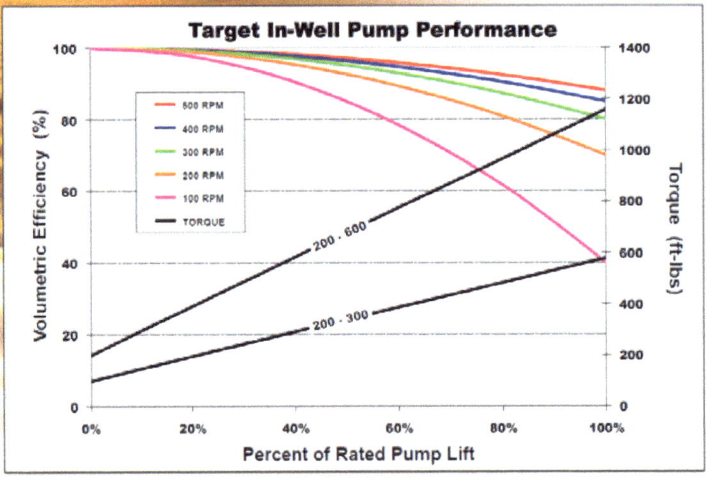

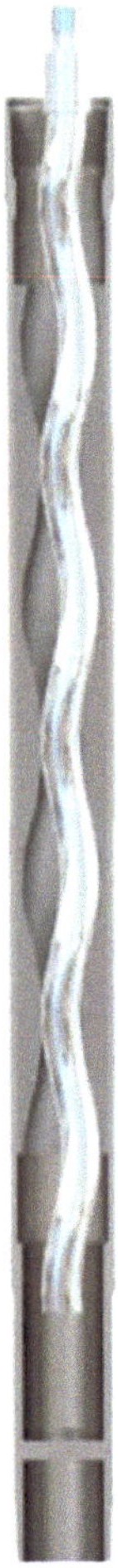

17 rue Ernest Laval BP 35 · 92173 Vanves Cedex France
Tel (33) 1 41 08 15 15 - Telex 634 129 F - Fax (33) 1 41 08 15 99
http://www.pcmpompes.com - Email:oil@pcmpompes.com

PCM MOINEAU OILFIELD

Specifications 1000 TP 600
ISO 1000/60

ROTOR				PUMP ASSEMBLY		
TOTAL LENGTH	9.010 m	29' 6.72"	No. OF STAGES (2)	11	
LENGTH OF HELIX	8.840 m	29'	HEAD CAPABILITY	600m	2000 ft
CREST TO CREST DIAM (1)	71.0 mm	2.79"	DISPLACEMENT	1458 cc	
HEAD DIAM (1)	57.5 mm	2.26"	CAPACITY PER RPM	2.10 m3/d	13.2 bpd
THREAD	19/16"API	1"1/8 rod	VOLUME AT 500RPM	1050 m3/d	6600 bpd
			O.D.	138mm	5.43"
STATOR			STOP BUSHING		
No. OF ELEMENTS	3		STAND-OFF LENGTH	0.3 m	1ft
LENGTH	8.490 m	27'10.2"			
O.D.	138 mm	5.43"			
THREADS	5" casing male				

(1) largest of the two is rotor O.D.
(2) Stage defined as equivalent to one pitch length of stator

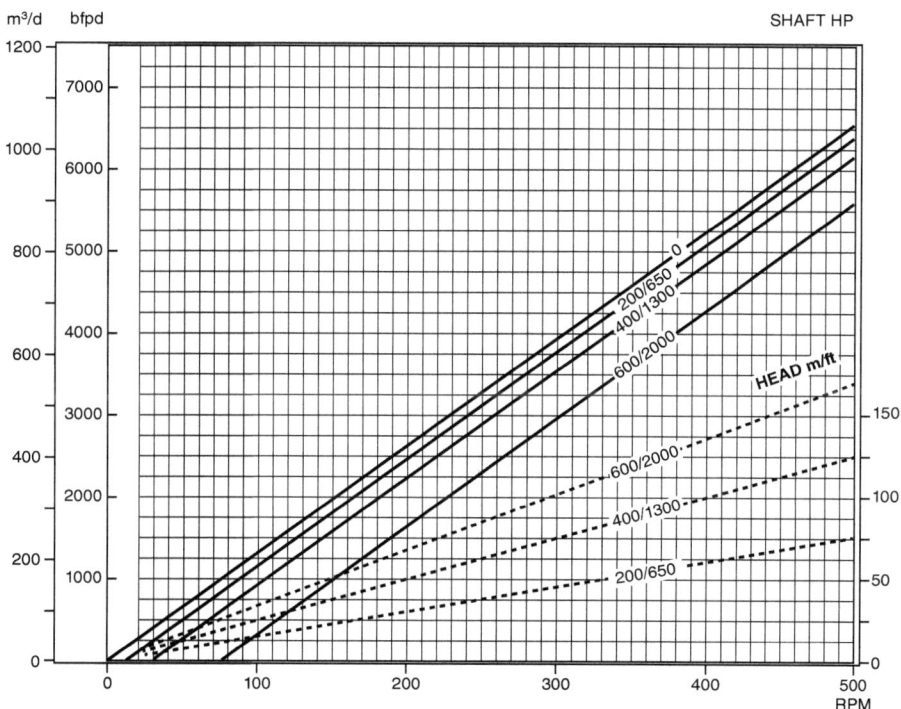

Figure 19-2

Example of pump specification sheet from PCM.
Model 1000TP600 (1000 m^3/day) at 500 rpm and 600 m pump head capability.
Source: *PCM catalogue*.

6.6 WELL DATA SHEET

It is a recommended practice to properly complete a data sheet (Figure 19-3) before any run-in-hole operation.

This data sheet includes the following items:

6.6.1 Well Situation

1. The Company name and address, the name of the contact.
2. The field location and the well identification.

6.6.2 Well Completion Data

1. Casing, liner and tubing characteristics.
2. Indication of open hole formation, or sand screens liner or perforation levels.
3. Tubing characteristics.
4. Well depth. Indication of the True Vertical Depth or the Total Measured.

Length if the well is deviated. A longitudinal well profile is requested, in order to appreciate the possible fitting inside the well curves, and the equipment to be operated.

6.6.3 Well Production Data

1. Mention the previous type of pump used in the well.
2. Desired depth and deviation for setting the pump in the well.
3. Oil static level in the well.
4. Dynamic level according to the flow rate, or indication of the productivity index.
5. Wellhead pressure according to the surface discharge conditions.

6. The current well capacity pumped or daily desired production rate of the flow of oil, water and gas considering the calculated GOR at inlet the Water/Oil Ratio and the total flow rate in m^3/day, barrel per day (bfpd) or gallons per minute (gpm).

7. The differential pressure between the pressure required at the pump discharge on surface and the pressure being supplied at the pump suction. Pressure in meters of H_2O, in bar (= 10 meters of H_2O), in pounds per square inch (PSI): (1 psi = 0.0689 bar, 1 bar = 14.5038 psi).

8. The maximum temperature at the pump inlet, temperature at which the product is to be pumped in Celsius (°C) or Fahrenheit (°F).

$$t°F = 9/5(t°C + 32), t°C = 5/9(t°F - 32)$$

Example: 0°C = 32°F, 104°C = 220°F, 149°C = 300°F, 204°C = 400°F

9. Temperature at surface.

Progressing Cavity Downhole Pump Data Sheet part 1

CUSTOMER		Company Name		Phone		Fax		Mail
	Date							
	Contact Name							
	Well Information		Well name/number		Field area			Country

WELL DATA

Well configuration		Vertical		Slant		Horizontal
	Directional survey	Wellbore Directional Survey (as drilled) for calculation of rod/tubing contact loads				

Down Hole Completion Details

Total vertical well depth, TVD		m			
Total well depth, measured		m		Temp Gradient	°C/100m
Perforations interval measured		m top depth	to		m bottom depth
Perforations interval TVD		m top depth	to		m bottom depth
Open Hole (from/ to)		m	to		m
Slotted liner (from/to)		m	to		m

Casing / Tubing Configuration

Casing String size&grade-weight		OD inch/mm			grade
		ID inch/mm			Thread
Tubing String size&grade-weight		OD inch/mm			grade
		ID inch/mm			Thread
Operating casing pressure		kPa		Operating tubing pressure	kPa
Anchor/Packers		type		Anchor/Packers Depth	m
Tubing length		m TVD		and	m measured
Pump Setting Depth		m TVD		and	m measured

PRODUCTION DATA

Productivity Index IP		$m^3/d/kPa$	Bottomhole Pressure	kPa
Fluid Level from surface		m TVD static	Target Fluid Level	m TVD dynamic
Current Production Rate		m^3/day	Target Production Rate	m^3/day
Gas Production Rate		m^3/day		
Temperature at pump inlet		°C	Well Head Temperature	°C
Temperature at Surface		°C		
Flowline Pressure		kPa	Casing Pressure	kPa

FLUIDS PROPERTIES / Oil &Gas Composition Analysis

Fluid	Fluid Temperature at pump inlet	°C	Total Fluid Specific Gravity	S.G
	Static Reservoir Pressure	kPa	**Producing Pressure**	kPa
	pH of Fluid		Bottom hole pressure at pump inlet (BHPI)	
Oil	**API Oil Gravity**	°API	Oil Specific Gravity	S.G
Water	Water Cut	%	Water Specific Gravity	S.G
	Water Salinity	ppm	Gas Specific Gravity	S.G
Gas	Gas/Oil Ratio	GOR m^3/m^3	Gas /Liquid ratio	GLR m^3/m^3
	Gas Bubble Point Pressure	kPa		
Solids	Sand Volume Produced	m^3/day	Sand Cut	%
Others	**H_2S concentration**	%	CO_2 concentration	%
	Mole percent Light Aromatics	Mole %	Chlorides	%

Fluid Viscosity / Temperature Correlation

Fluid Viscosity at pump setting depth	centipoises (cp)	at	T°C
Fluid Viscosity at surface	centipoises (cp)	at	T°C
Fluid Viscosity/Temperature	centipoises (cp)	at	T°C
Fluid Viscosity/Temperature	centipoises (cp)	at	T°C

Chemicals Treatment Programmes

	Yes		No
If Yes Describe	Corrosion Inhibitors		Scale Inhibitors
Other (Please Specify)			

Progressing Cavity Downhole Pump Data Sheet part 2

Well Head Description

Flow Line		type		size	Thread
Wellhead connection		type		size	Thread
Tubing Head type / size		Flange		size	Thread

PCP Drive Head

Power Source	Electric		Hydraulic	Diesel / Gas Engine
	voltage	frequency		
Drive Head Type	Belt & Sheave Ratio		Gear Box Ratio	Direct Drive / Hydraulic Drive

Electric Drive Head

	Type		Voltage
Power rating	hp		kVA
Motor Speed rating	rpm	at nominal frequency	Hertz
Variable Frequency Speed	rpm at F min Hz		rpm at F max Hz

PCP Rods Drive

Polished Rod size&grade		inch/mm	and		grade
Drive string type	Solid	Coiled	Hollow	Continuous	
Drive string rods size		OD inch/mm		coupling OD inch/mm	
		Thread type	and		grade
Pony size-grade		inch/mm	and		grade
Rod Guides/Scrapers		type			
Torque anchor		type		Gas Separator	type

TARGET DAILY PRODUCTION / DAILY PRODUCTION

TARGET DAILY PRODUCTION	Fluid Rate	DAILY PRODUCTION	Fluid Rate
Oil	m^3/day	Oil	m^3/day
Water	m^3/day	Water	m^3/day
Gas	m^3/day	Gas	m^3/day

Figure 19-3

PCP data sheet well completion and production.
Source: *IFP Energies nouvelles*.

6.6.4 Production Fluid Data

A good knowledge of the physicochemical characteristics of the oil, water and gas fluids helps to:

1. Determine the pumped volumes of oil, water and gas considering the calculated GOR at inlet.
2. Select a compatible elastomer with the pumped fluid.

The supplier should be given a fluid sample of the well in order to test elastomers.

The essential data provided includes:

1. **Fluid characteristics:** Oil, Water, Gas.

Oil:

- Specific Gravity = Ratio of the density of the product-to-water at the pumping temperature.
- Viscosity at the pumping temperature.
  The SI viscosity units is the Pascal.second (Pa.s), the CGS viscosity unit is the poise (P), (1 Poise = 0.1 Pa.s, 1 cP = 10^{-3} Pa.s = 0.001 Pa.s).
- Aromatic content.
- Mole percent of H_2S, CO_2, chlorides (in %).

Water:

- Gravity.
- Salt and other solids content.
- Water/Oil ratio.

Gas:

- Standard GOR.
- Specific gravity.
- Bubble point.

2. **Nature of the fluid**, corrosion – abrasion.

- **Corrosion:** The pH value (hydrogenation activity) of the product should be known in order to select suitable construction materials for the pump.
- A pH of 0-7 is acid, pH of 7 is neutral, and a pH of 7-14 is alkaline.
- How corrosive is the liquid? Will it lubricate the solids and reduce abrasion?
- **Abrasion:** The wear characteristics based on the abrasive nature of the solids in the fluid.
- **Concentration of solids:** The ratio (%) of solid-to-liquid on a volume basis contained in the product. This ratio will usually determine how the characteristics of the solids influence the slurry as a whole.
- **Size and Nature of the solids:** The size of the largest particles the pump must handle. Are they soft or hard, light or dense, round or jagged, fibrous or stingy, abrasive or corrosive?

6.6.5 Well and Fluids Characteristics Influence Pump Performance

1. Pressure

- Increasing differential pressure increases pump slippage, i.e. the leakage between the chambers.
- Higher Pressures require additional stator/rotor stages.
- Higher pressures and slip have an accelerating effect on the rate of abrasive wear.

2. Temperature

- Higher temperatures affect the choice of elastomeric materials.
- Higher temperatures require the use of undersize rotors.

3. Viscosity

- As viscosity increases, the pump slip will decrease.
- As viscosity increases, the maximum allowable speed decrease.
- As viscosity increases, the maximum allowable pressure per stator/rotor stage increases.

4. Concentration of the Solids

- Mixed liquids and microscopic solids usually produce a non-Newtonian liquid.
- Higher solids in suspension content decreases maximum allowable speed and pressure per stage.

5. Size and nature of the Solids

- Soft solids (abrasive) reduces the maximum allowable pump speeds.
- Pumps must have sufficient cavity size to pass the largest hard particle in the product.

6. Abrasion

- The more abrasive is the product, the lower is the pump speed to operate safely.
- The amount of wear on an abrasive application is closely proportional to the speed squared.

6.6.6 Operational Specifications

The operator may require minimum production rate conditions and define a pump setting depth in the well.

According to the *in situ* conditions, a specific energy supply system may be chosen: electric, hydraulic, gas. Depending on the production schedules, the motor transmission may be at fixed or at variable speed (Chapter 9: Drive head).

6.6.7 Physicochemical Treatment Programme

A chemical treatment may be provided by the operator, either at the wellhead, or at the bottomhole. A completion modification is then possible. Also, it may be necessary to check the elastomer behaviour.

6.6.8 PCP Data Sheet

An accurate data sheet is required to comprehensively evaluate and design a complete PCP' system that will provide satisfactory results.

See PCP data sheet Figure 19-3.

CHAPTER 7

Presence of Gas at the Pump Inlet

The Progressing Cavity Pump can pump oil and gas mixtures without difficulty. But as they are positive-displacement pumps, the volume of gas pumped means less oil production.

The efficiency of PCP is drastically reduced if the liquid being pumped contains large volumes of gas.

7.1 VERTICAL OR SLIGHTLY DEVIATED WELL

It is wise to place the pump below the bubble point which – in many cases – is the level of the reservoir. An almost automatic separation occurs at the reservoir level: the gas goes up to the surface, whereas the oil flows downwards so as to be sucked up by the pump.

7.2 HIGHLY DEVIATED OR HORIZONTAL WELL

In this case, the pump is generally located above the pay zone. It is thus subjected to a two-phase flow at its inlet.

This two-phase flow may be (Figs. 20-1 and 20-2):
- Permanent and homogeneous (gas bubbles dispersed in oil).
- Stratified (superposition of gas and oil).
- Intermittent (gas pockets topping a liquid film or liquid slugs with gas bubbles).

7.2.1 Permanent and Homogeneous Flow

For a given flow rate, the highest possible submergence pressure has to be created so as to generate a low GOR at the inlet.

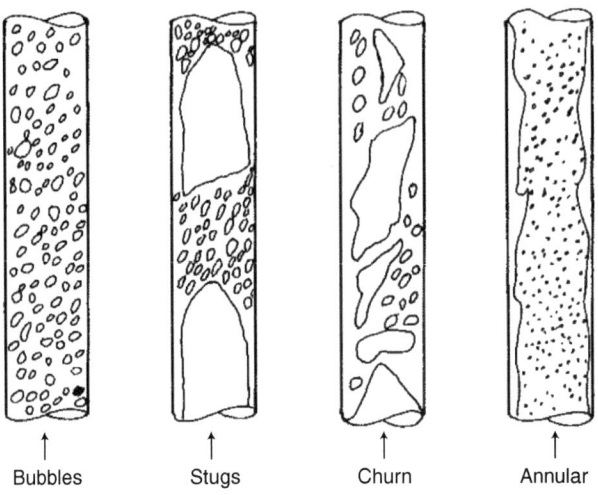

Figure 20-1

Flow pattern in vertical wells.
Source: *A.E. Dukler. University of Houston.*

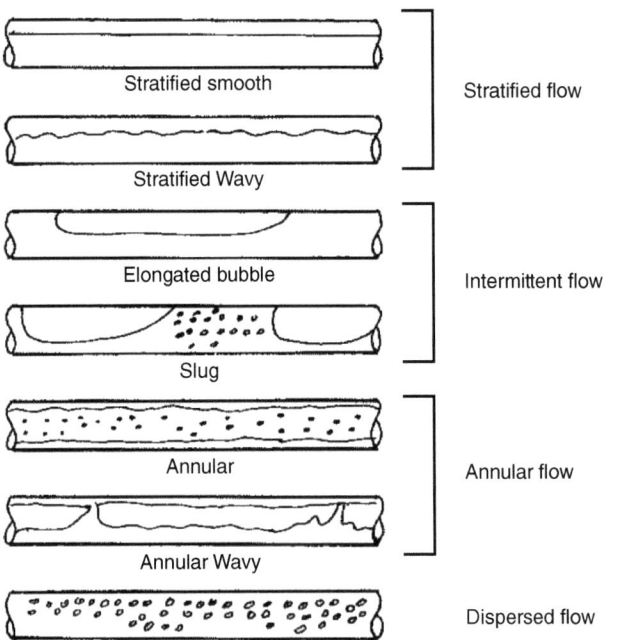

Figure 20-2

Flow pattern in horizontal pipes.
Source: *A.E. Dukler, University of Houston.*

7.2.2 Stratified or Intermittent Flow

An effluent with a widely varying GOR is likely to damage pumping equipment.

7.3 A "NATURAL" GAS SEPARATOR

The best technique for separating gas and production oil is to set the orifices of the inlet tubing below the reservoir. The natural separation of gas and liquid is relatively well achieved with flow speeds of below 15 cm/s in the annulus. The pump inlet orifices should be located at 5 meters minimum under the gaseous formation so a free separation of gas and oil can rise up into the casing/tubing annulus. These values are depending on the pumped flow rate.

7.4 STATIC GAS SEPARATOR

When the above conditions cannot be obtained a static two phase liquid-gas separator is recommended at the pump inlet.

The basic principle of such a separator is the creation of a baffle crossing between two concentric tubes. Figure 21 shows the equipment comprising the pump included inside the separator. A small diameter lateral pipe achieves the possible injection of a diluent reducing the oil viscosity at the pump inlet.

Research on multiphase flows has demonstrated that large gas bubbles are carried along in the fluid. When the separator is concentric with the casing on Figure 21, the gas distribution is uniformly distributed throughout the annulus space.

7.4.1 Static Continuous Downhole Centrifugal Gas Separator

To improve these separation conditions in vertical and deviated wells, a specific PCP downhole static continuous flow gas separator is provided which can be attached to the suction part of the PCP. It removes gas from liquid being pumped prior to the liquid entering the pump inlet (Fig. 22).

The downhole static continuous flow gas separator is used in conjunction with all types of progressing cavity pumps in vertical to deviated wells. It must be installed in the string below the pump and above the perforations. If it is desired or necessary for the fluid intake to be below the perforations, tubing may be run below the gas separator (tail joints).

The gas separator uses the centrifugal force principle to separate the gas from the produced liquids before they enter the PCP. The liquid-gas mixtures enter through multi-intake ports to flow in an annular path through the auger chamber, which subjects the fluid to centrifugal forces from the inlet to the outlet end. The more dense materials (e.g. water and oil and sand)

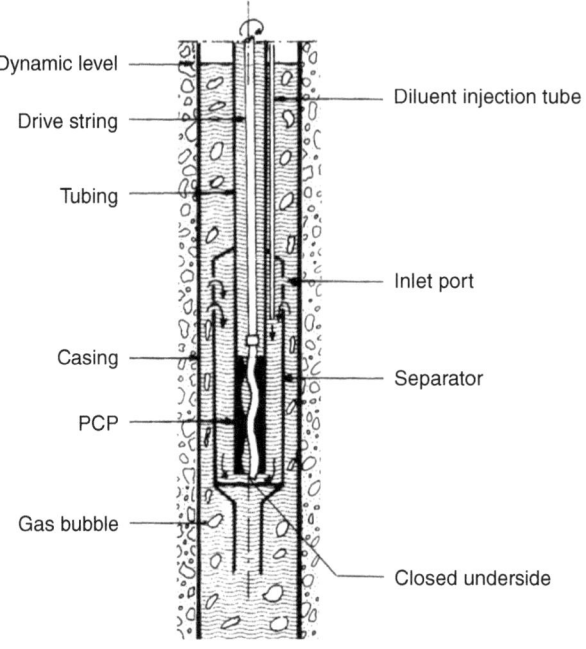

Figure 21

Static gas separator.
Source: *IFP/Horwell*.

will be forced to the outside of the chamber while the gas will remain close to the middle. As the fluids reach the top of the auger chamber, liquid is removed *via* outlet ports (around the perimeter of the top sub) which lead to the pump suction itself. Meanwhile the gas contents in the central region reach the top sub of the auger chamber from which it is removed *via* separate central gas outlet ports to migrate up to the casing-tubing space annulus.

With significant to high gas content in the fluids at the PCP inlet, many problems will occur such as:

- Low PCP efficiency.
- Too high torque due to poor lubrication with sand content and gas.
- Short PCP working life. The high gas content can cause the stator to burn up and break apart due to friction without rotor-stator lubrication.
- Elastomer swell and blister.
  Exposed to excessive levels of aromatic conventional elastomer swell; and in high-gas contents, they blister under explosive decompression conditions.
  Over time when the stator elastomer is exposed under high pressure gas CO_2, H_2S and aromatic, gases enter the stator elastomer by solution and diffusion processes to cause early deterioration. When the PCP is pulled out of the well, the outside pressure is removed rapidly. The elastomer becomes supersaturated and gas comes quickly out of solution, causing blistering and fractures it (so-called explosive decompression).

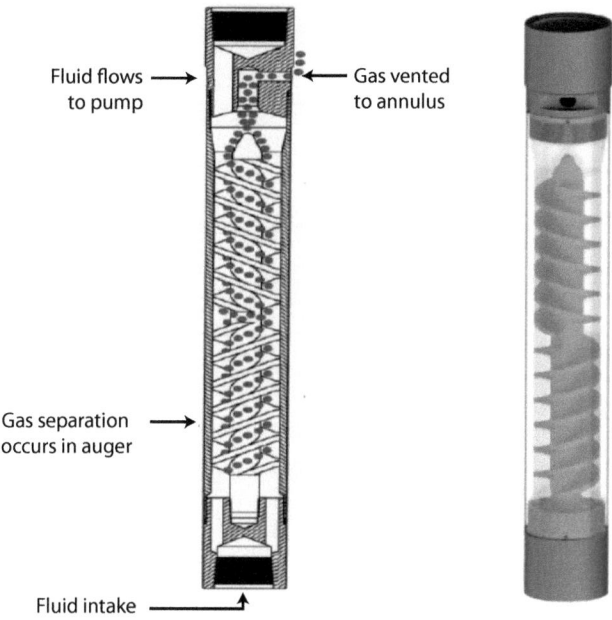

Figure 22

Centrifugal gas separators.
Source: *CANAM, Evolution Oil tools catalogue*.

The main advantages in and reasons for using gas separators to eliminate the associated gases before the pump intake are:

– As PCP performance or pump efficiency is increased, drive string torque is reduced.
– A longer run life for the PCP stator elastomer, less early elastomer deterioration.
– This results in lower operating costs.
– The longer PCP run life in gassy conditions minimizes the need for well intervention procedures, as well as the associated costs; lower power consumption further reduces costs.

Gas separators located below torque anchor are available in different configurations for light oil > 24° API and for more heavy oil < 24°API and sizes from manufacturers and suppliers as: CANAM, Evolution Oil Tools, PCP Oil Tools.

7.4.2 System Avoiding Gas Suction in the Horizontal Wells

The setting of a static separator identical to the vertical well is recommended, and as described in the previous paragraph, would have a low efficiency because the level difference between the inlet and outlet separator is shallow. Also, when the effluent dynamic level is low compared to the inclined part of the well, the setting of the pump and separator is almost impossible in a steeply inclined section.

In many cases, oil and gas constitute a stratified flow into the horizontal well (Fig. 20-2). In a typical situation, the horizontal section of the wellbore is only half full of liquid thus allowing the pump to draw large amounts of gas that will affect the efficiency and run life of the pump.

To run in such conditions, the pump is often set in the deviated part of the well, but under the dynamic level, and extended with a production tube (tubing), up to the horizontal drain.

To suck up the reservoir oil along the lower generator of the horizontal well.

At the tubing extremity:

1. A suction orifice is fitted in a lateral manner considering the suction means and operating in one direction.

2. The orifice direction is oriented to approximately face the horizontal well's lower generator. An original system including magnets was designed for this purpose.

Figure 23 illustrates this IFP patented "method and device for producing by pumping in horizontal, drain hole" H.Cholet, C.Wittrisch US 5,829,529.

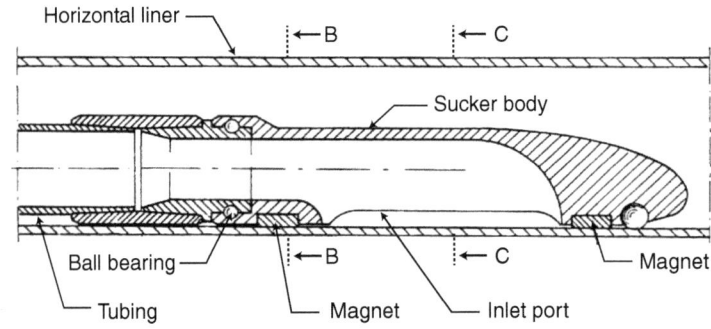

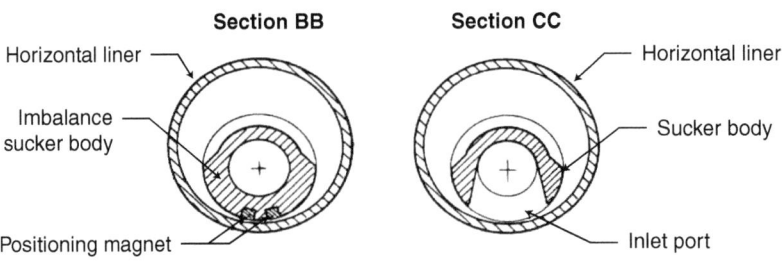

Figure 23

Gas separator systems for horizontal wells.
Source: *IFP Energies nouvelles.*

The Eccentric Intake Sub from Manufacturers

The eccentric intake sub is a specific sub run end of the tubing instead of traditional open ended tubing. It is efficient from 71 degrees deviation to the horizontal section of a well. The eccentric intake sub is a weighted eccentric mandrel that will always orient the intake to draw from the bottom of the horizontal wellbore. This will allow the pump to draw more liquid and less free gas.

Manufacturers and suppliers are: Coteco, Evolution Oil Tools, Premium Artificial Lift system, Masteroil Tool.

7.5 GOR CALCULATION AT THE PUMP INLET

It is essential to determine the gas effect on the fluid volume in order to accurately select the pump model, and consequently to know the actual GOR at the pump inlet.

If the solution gas/oil ratio (R_s), the gas volume factor (B_g), and the formation volume factor (B_0) are not available from reservoir data, they must be calculated. When these three values are known, the volume of oil, water and gas can be determined and percentages of each calculated. The formulas are provided in the metric system.

7.5.1 Solution Gas/Oil Ratio

$$R_s = 0.13422 \times Y_g \left[P_b \times \frac{10^{0.0125 \times °API}}{10^{0.00091 \times (1.8T+32)}} \right]^{1.2048}$$

where:
- Y_g specific gravity of gas.
- P_b bubble point pressure.
- T temperature at pump inlet (in °C).
- R_s must be corrected "R_s(corr.)" when the submergence level is below the bubble point.

$$R_s(\text{corr.}) = R_s \times \text{factor}$$

The factor is taken from Figure 24, by locating the ratio of "Submergence pressure/Bubble point pressure" on the graph and determining the corresponding ratio of "GOR remaining in solution/Solution GOR at bubble point", which is multiplied by the calculated R_s to get R_s(corr).

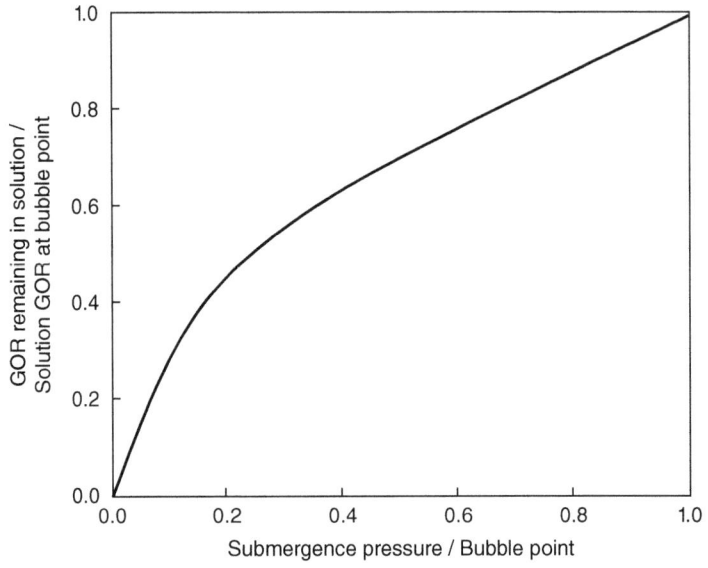

Figure 24

Non-dimensional gas liberation curve.
Source: *Nabla Corporation*.

7.5.2 Gas Volume Factor

$$B_g = 0.00378 \times \frac{ZT}{P}$$

where:

- Z gas compressibility factor (0.81 to 0.91).
- T bottom-hole temperature (in °Kelvin (°C + 273)).
- P submergence pressure (in bar).

The gas volume factor B_g is expressed in m³/m³.

7.5.3 Formation Volume Factor

$$B_0 = 0.972 + 0.000147 \times F^{1.175}$$

where:

$$F = 5.61 \times R_s(\text{corr.}) \sqrt{\frac{Y_g}{Y_0}} + 1.25(1.8T + 32)$$

where:

Y_g specific gravity of gas.
Y_0 specific gravity of oil.
T temperature at the pump inlet (in °C).
$R_s(\text{corr.})$ is defined from the figure 24 (5.1.1 above).

The formation volume factor B_0, represents the ratio of an oil volume in the formation compared to a stock tank volume.

From R_s, B_g, B_0, the percentage of gas by volume can be calculated.

7.5.4 Total Volume of Fluids

Total gas volume (free and in solution)

$$V_{gt} = \text{Stock Oil Volume} \times \text{GOR}_{standard}$$

$\text{GOR}_{standard}$ is the gas/oil ratio measured at surface (m^3/m^3).

Gas volume in solution at submergence pressure

$$V_{gs} = \text{Stock Oil Volume} \times R_s(\text{corr.})$$

Free gas volume

$$V_{gl} = V_{gt} - V_{gs}$$

Oil volume in the formation

$$V_0 = \text{Stock oil volume} \times B_0$$

Gas volume in the formation

$$V_g = V_{gl} \times B_g$$

Water volume into the formation V_w

Total volume

$$V_t = V_0 + V_g + V_w$$

So, for a certain oil volume measured in the stock tank, the pump must be selected to lift the calculated volume V_t.

7.6 AN APPLICATION EXAMPLE

These relationships enable the operator to know what fluid volume (oil + water) will be measured in the stock tank, according to the pump submergence level, and the GLR (Gas/Liquid Ratio) at the pump inlet. A pumping flow rate will be clearly higher than the stock tank flow rate.

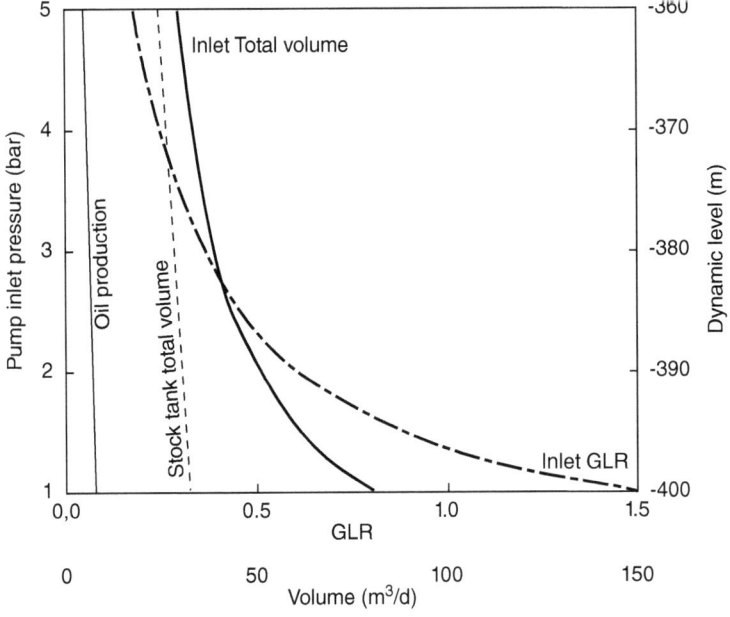

Figure 25
Relation between fluid level and PCP performance.
Example of a Canadian well.
Source: *C.S. Resources/IFP Energies nouvelles.*

Figure 25 is an example of a well producing in Canada, which show, according to the submergence level, the increase in the pumped fluid volume ratio (oil + water) in the formation, compared with the total volume (oil + water + gas). With two-phase pumping, a variation of the submergence level highly influences the positive displacement pump efficiency.

Pumping with a high GLR at the pump inlet would reduce the life span of the elastomer.

CHAPTER 8

Drive Rods

Generally, PCPs are driven from the surface by the same type of sucker solid rods as those used for beam pumps, but to drive the PCP rotor requires rods with high torque stress, named "drive rods".

Specific drive rods are described in 8.4, 8.5, 8.6 as: small diameter sections of tubing, hollow rods, continuous rods or coiled tubing.

8.1 SOLID DRIVE ROD CHARACTERISTICS

PCP drive rods range from 3/4" (19 mm) to 1 1/2" (38.1 mm) body size, with lengths of 25 or 30 feet (7.6 m or 9.1 m) and shorter pony rods.

Different API and proprietary steel grades are available from manufacturers to drive PCP rods string. Rod performance and selections are determined by steel grades, manufacturing

Figure 26

Solid drive rods.
Source: *Tennaris catalogue*.

processes, heat treatments, loads, yield strengths, torque limits and corrosive resistance levels in the wells.

Referring manufacturers include: Tennaris, Norris, Alberta Oil Tool, Canam, Cameron-Camrod.

Drive rod API standard steel grades used are:

- **Grade C and D (Carbon) AISI 1530 M**: to drive rods with low to **medium loads** for **non-corrosive** or effectively inhibited well fluids. Minimum yield strength 60 kpsi – 413 MPa, ultimate strength 115 kpsi – 792 MPa. (Tennaris)
- **Grade K AISI 4621 M**: to drive rods with low and **medium loads** in **corrosive** wells fluids where inhibition is recommended. Minimum yield strength 60 kpsi – 413 MPa, ultimate strength 115 kpsi – 792 MPa. (Tennaris).
- **Grade D (Carbon) AISI 1541 VM** Carbon-Manganese Alloy steel, to drive rods with **medium** loads and for **non-corrosive** or in effectively inhibited wells fluids. Minimum yield strength 115 kpsi – 793 MPa, ultimate strength 140 kpsi – 965 MPa (Canam).
- **Grade D (Alloy) AISI 4142 M** Chromium-Molybdenum Alloy steel, to drive rods with medium to **high loads** and for non- or **medium-corrosive** well fluids **effectively inhibited** again corrosion attack. Minimum yield strength 100 kpsi – 690 MPa, ultimate strength 140 kpsi – 965 MPa. (Tennaris, Weatherford, Canam).
- **Grade KD AISI 4320 M**: Nickel-Chromium-Molybdenum Alloy steel to drive rods with **medium loads** and for **corrosive** wells fluids where inhibiting is recommended. Minimum yield strength 85 kpsi – 585 MPa, ultimate strength 140 kpsi – 965 MPa. (Tennaris).
- **Grade D Special alloy AISI 4330 M**: Nickel-Chromium-Molybdenum Alloy steel, designed to maximise corrosion and fatigue to drive rods with medium to **high loads** and with **corrosive** well fluids where inhibiting is recommended. Minimum yield strength 100 kpsi – 690 MPa, ultimate strength 140 kpsi – 965 MPa. (Tennaris).
- **Grade HD Special alloy AISI 4332 SRX**: Nickel-Chromium-Molybdenum Alloy steel, Minimum yield strength 115 kpsi, 793 MPa, ultimate strength 140 to 150 kpsi, 965 to 1,034 MPa (Weatherford)

Other proprietary steel Grades and specially designed rods to support high torques, high loads for deep wells with high flow rates are available from manufacturers.

(1 kpsi = 6.89 MPa, 1 MPa = 0.145 kpsi)

The solid rods used have the following tensile strengths:

Drive Rods		Minimum strength		Ultimate strength		Corrosion resistance	If inhibited
Grade	AISI	kpsi	Mpa	kpsi	Mpa		
C/D Carbon	1530M	60	413	115	792	NON	YES
K	4621M	60	413	115	792	YES	
D Carbon	1541	115	792	140	965	NON	YES
D Alloy	4142	100	690	140	965	Medium	YES
KD Alloy	4320M	85	585	140	965	YES	Recommended
D Special Alloy	4330	100	690	140	965	YES	
HD Special Alloy	4332	115	793	140/150	965/1034		

The characteristics of the drive rods depend on:
1. The tubing diameter.
2. The mechanical stresses generated by:
   - The axial load of the rods.
   - The thrust generated by the head rating of the pump.
   - The mechanical resistant torque.
   - The resistant torque due to the viscosity of the effluent in the tubing.

On the surface, a Progressing Cavity Pump is characterized by its drive system which includes:
- A drive head.
- A motor.
- A rotating speed reducer system.
- Polished rods and shorter pony rods.

A code identifies the installed drive head.

8.1.1 Influence on the Tubing Diameter

The pressure loss in tubing depends on the tubing size and the drive rod size.

Pressure loss improvements have been implemented by reducing the rod coupling size *versus* the ID tubing size. "Slim hole couplings" are available from manufacturers to increase the tubing flow area.

Based on the tubing diameter, the drive rods which may be used are the following.

Tubing outside diameter	Drive Rod Body	Full size Pin	Slim hole Pin
2 3/8"	3/4"	7/8"	3/4"
2 7/8"	1"	1""	7/8"
3-1/2"	1-1/4"	1 1/8"	1"
	1-1/2"	1 1/8"	

8.1.2 Drive Rod Characteristics

Drive rod characteristics from manufacturers are:

Drive Rods Body Nominal diameter		Outside Diameter of coupling	Rod Section	Approx Weight in air	Coupling Regular OD	API Thread Size
(inches)	(mm)	(inches)	(cm^2)	(kg/m)	(inches)	(inches)
3/4	19.0	1-5/8	2.85	2.37	1-5/8	1-1/6
7/8	22.2	1-13/16	3.88	3.17	1-13/16	1-3/16
1	25.4	2-3/16	5.07	4.20	2-3/16	1-3/8
1-1/8	28.6	2-3/8	6.41	5.36	2-3/8	1-9/16
1-1/4	31.75		7.91			
1-1/2	38.1		11.39			

8.1.3 Maximum Service Rod Torques

The maximum service torques for drive rods and pony rods referring Alberta Oil Tool catalogue are:

Drive Rods Size	Grade D Alloy AISI A-4142-m		Grade D Special Alloy AISI A 4330-M		Special High Strength Grade AISI 4138-M AISI 4330-M	
inches	ft.lbf	daN.m	ft.lbf	daN.m	ft.lbf	daN.m
1	1,100	149	1,110	151	1,200	163
1-1/4	2,000	271	2,100	285	2,500	339
1-1/2	3,000	407	3,150	427	3,750	509

(1 ft.lbf = 1 lbf.ft = 1.356 N.m = 0.1356 daN.m = 0.13825 kgf.m

1 daN.m = 7.37 ft.lbf, 1 kgf.m = 7.23 ft.lbf)

8.2 STRESSES IN THE DRIVE STRING

These stresses are due to the following.

8.2.1 Weight of Solid Rods and Hollow Rods

Solid and hollow rods are currently used to drive the down hole PCP from the surface.

They have the following characteristics:

Nominal diameter (inches)	Nominal diameter (mm)	Rod section (cm^2)	Weight in air (kg/m)
SOLID RODS			
3/4	19.05	2.85	2.37
7/8	22.23	3.88	3.26
1	25.40	5.07	4.20
1-1/8	28.58	6.41	5.36
1-1/4	31.75	7.91	6.70
1-1/2	38.10	11.39	9.60
HOLLOW RODS			
1.89 flush	48	18.08	6.00
1.69	42	13.84	4.90
1.89	48	18.08	6.10

For calculating the weight of the rods in operation, the buoyancy generated by the fluid contained in the production tubing should be taken into account.

8.2.2 Thrust Generated by the Head Rating of the Pump

Depending on the pump model chosen, the load on the thrust bearing of the drive head (see chapter 3-1-5) is:

$$F_b = \frac{\pi \times \Delta P \times (2E + D)^2}{4}$$

8.2.3 Mechanical Resistant Torque

The mechanical operational resistant torque Γ_m defined in chapter 3-1-4 is:

$$\Gamma_{m(daNm)} = 1.63 \times V_{(cm^3)} \times \Delta P_{(kPa)} \times 10^{-5} \times \rho^{-1}$$

V being the pump cylinder capacity of the pump and ρ its efficiency (for the evaluation, $\rho = 0.7$).

8.2.4 Resistant Torque by the Effluent Viscosity in the Tubing

See chapter 4-3-7. This is determined by the following relationship (expressed in practical units).

$$\Gamma_v = 0.165 \times 10^{-8} \times \mu_f \times L \times N \times \frac{d^3}{D-d} \times \frac{1}{\ln\frac{\mu_s}{\mu_f}} \left(\frac{\mu_s}{\mu_f} - 1 \right)$$

where:
- Γ_v Resistant torque (in daN.m).
- μ_f Viscosity of the effluent at the inlet (in cP).
- μ_s Viscosity of the effluent at the surface (in cP).
- L Length of the tubing (in m).
- N Rotating speed (in rpm).
- d Rod diameter (in cm).
- D Inside diameter of the tubing (in cm).

8.2.5 Total Stresses in the Drive Strings

The drive strings are subjected to:

1. A tensile stress F generated by their own weight F_p in the fluid, and a downward thrust due to the head rating of the pump:

$$F = F_p + F_b$$

A torsion Γ generated by the resistant torques:

$$\Gamma = \Gamma_m + \Gamma_v$$

These stresses are different from those generated with conventional beam pumps.

The stress σ_t in the drive string is the resultant value from these values, that is:

$$\sigma_t = \frac{0.4}{\pi d^3} \sqrt{F^2 d^2 + 64\Gamma^2 \times 10^6}$$

where:

σ_t resultant stress (in MPa)
d diameter of the drive string (in mm)
F axial load in the drive string (in daN)
Γ resistant torque (in daN.m).

8.3 TUBINGS AND DRIVE STRING

Consideration should be given concerning the choice and the implementation of the tubing and the drive strings to ensure perfect harmonization with the actual pump.

8.3.1 Tubing and Centralizers

Tubing Precaution

The rotor-lower rod connection, located above the stator, undergoes the eccentric movement of the rotor. Therefore, care must thus be taken with the dimensions of the first joint fastened above the stator to achieve this eccentric movement. Manufacturers draw up a table relating to the compatibility between pump/drive string/tubing for each type of pump.

In the event of high rotating speeds, the eccentric movement generates vibrations. The combination of the vibrations and the clockwise rotation tends to loosen the connections. To avoid such mishaps, the tubing connections should be tightened to the maximum torque recommended by API standard.

As a further precaution, and if the casing diameter permits, tubing centralizers may be used to absorb the vibrations and to centralize the pump inside the casing. Depending on the pump length, one or two centralizers are installed on the pump and another on the stator-tubing joint.

The proper positioning of the tubing in the well is achieved by means of a tension packer or Torque anchor (No-Turn-Tool).

8.3.2 Drive Rod String and NON Rotating Centralizers

Apart from the problems linked to the stress in the rod drive strings generated by the axial load, the resistant torque and the oil viscosity, it is also very important to avoid friction of the coupling of two successive drive strings against the inner wall of the tubing, particularly in deviated wells.

At pump level, rod-tubing wear may take place due to the eccentric motion of the rotor.

At high rotational speeds, the eccentric motion of the rotor from being transmitted to the rod string tends to belly out, thereby chafing against the lower tubing joint. To prevent this occurrence it is advisable to use a 12 ft pony rod on top of the rotor and to fit one non-rotating centralizer twelve feet above the rotor head plus one on top of each of the two to four full drive rods. **However, never use a short pony on top of the rotor and never fit it with a rod centralizer.**

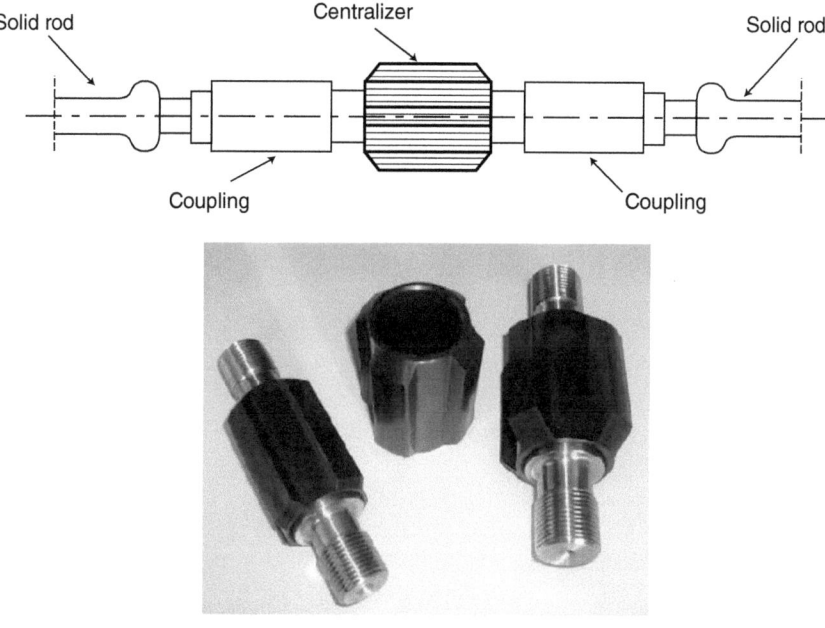

Figure 27

Drive rod and NON-rotating centralizers.
Source: *PCM and Canam*.

Where the surface rod string vibration is transmitted to the polished rod, which reduces the life of the seal or stuffing box, one centralizer should be placed at the bottom of the polished rod and one at the adjacent rod.

In deviated wells, non-rotating centralizers in the drive rod strings act as bearings in the inclined parts of the tubing.

Manufactures recommend a rod guide programme to select the right centralizer, such as type, size, material and spacing, wear rate and wear zones depending on well completion, well bore inclination, dog-leg severity, rod drive and pump characteristics, well fluids and others criteria.

8.3.3 Case of Deviated and Horizontal Wells

Deviated and horizontal wells run by drive rods strings are sensitive to wearing problems in contact with tubing and drive strings. The wear depends on:

- The drive string shape.
- The applied load of the string on the tubing.
- The effluent produced (and eventually the presence of any sand).
- The rotating speed of the drive string.

The magnitude of the contact load of the string of the tubing depends on the well curve and on the weight of the drive string. In the case of the usual drive rods, the contact forces are concentrated at the coupling level, or the fitted centralizers. Yet, in the case of a continuous drive string, the contact forces are distributed over a large distance, and are then much weaker by length unity.

Experience has shown that wear increases linearly with contact force.

Several types of centralizers are offered to operators. They actually limit the wear between drive strings and tubing, but may generate extra pressure drops in case of viscous oil production.

8.3.4 Contact Between the Tubing and the Drive Strings

Protection Against Wearing

The utilization of PCP in horizontal or highly deviated wells clearly showed frequent wear problems where drive strings and tubing make contact.

Without effective protection this leads to tubing piercing, broken strings, production drops, and well equipment being returned. An increase in pump performance requires effective protection against frictional wear.

8.3.4.1 Concept of a Centralizer for PCPs

Generally, in highly deviated wells, the drive string is leant against a tubing generatrix and generates a concentration of forces at the rod coupling level. The setting of fixed centralizers similar to those used with beam pumping is not effective enough, because of their concept design which is not scheduled for rotating. They generate an increase in resistant torque and wear of their wings happens quickly.

Consequently, a non-rotating centralizer has been specifically designed to be fitted on the PCP drive strings. It should be fitted at the junction of two drive strings and behaves as a bearing, As shown in Figure 27, the centralizer axis is interdependent of the drive strings, whereas the straight blades of the centralizer are against the tubing, improving the guiding and the stability of the drive string rotation. With this operating principle, there is non-rotating contact between the drive strings and the tubings. Then it is preferable to fit straight blade centralizers, not helical, in order to achieve better support against the tubing generatrix. The centralizers are designed with high resistant plastic material, generating a low frictional coefficient between the metallic coupling and the centralizer.

8.3.4.2 Centralizer Distribution Along the Drive Strings

The drive string undergoes a tensile stress which induces contact of the drive string against the internal generator of the tubing (Fig. 28). Centralizers are then sized in a way that should avoid any contact between two couplings.

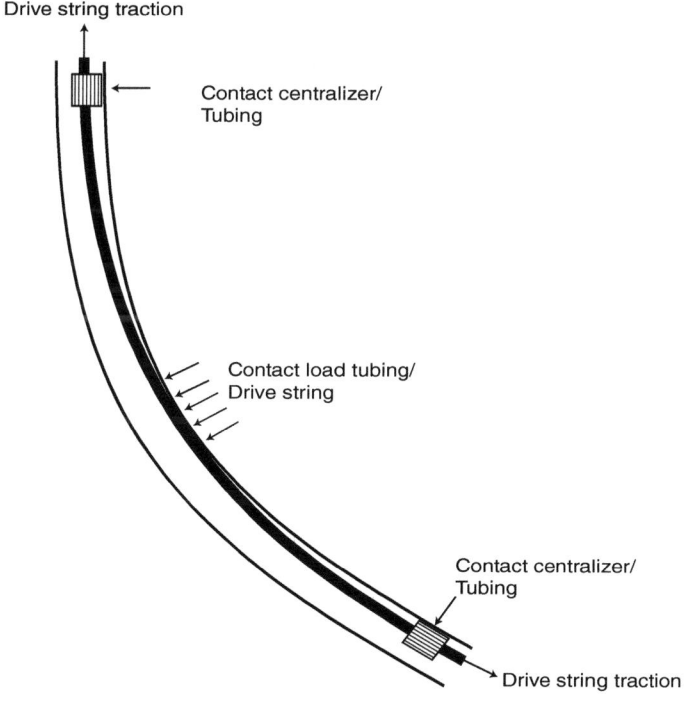

Figure 28

Drive string/Tubing contact geometry.
Source: *IFP Energies nouvelles*.

Each PCP distributor has a calculation programme which can define the centralizer location on the drive string, depending on:

- The tubing diameter.
- The drive string diameter.
- The tensile stress on the drive string.
- The well profile and the bending radius.
- The rotating speed.
- The physicochemical characteristics of the pumped fluid.

In a vertical well, 5 centralizers may be enough, but in a highly deviated well, a centralizer can be fitted on each drive rod. Pony rods can also be set to increase the number of centralizers.

In order to ensure reasonable wear rates for centralizers, the leaning stress on the internal wall of the tubing may be limited. In the event of low quantities of sand being present, the load contact centralizer/tubing should not exceed 500 N. If the sand rate increases, this value may be reduced.

8.3.4.3 Example of a Pump Equipped with Centralizers

Figure 29 shows the distribution of centralizers along the drive string, according to the stresses applied on the tubing.

This graph clearly shows that the lateral stresses are very significant at the deviation origin and become rather weak when the angle of inclination remains almost constant.

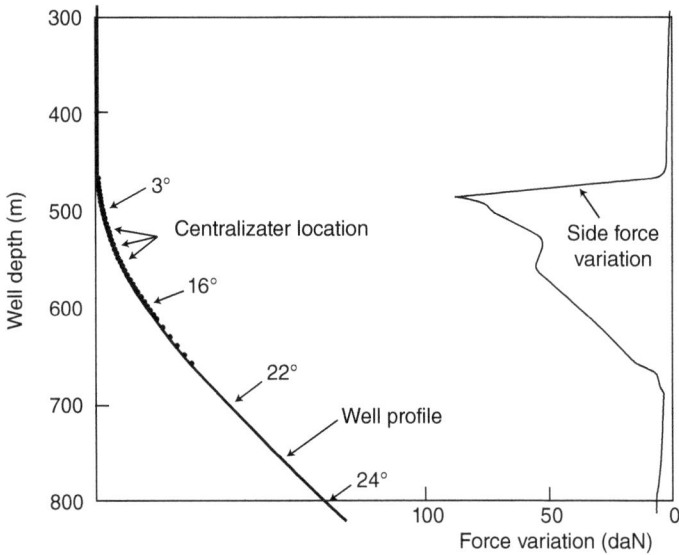

Figure 29

Example of centralizer location in relation to side force *versus* well profile.
Source: *PCM*.

Centralizer Selection and Advantages

Depending on well completion, well bore inclination, dog-leg severity, rod drive and pump characteristics, well fluids and other criteria, manufacturer rod guide programmes recommend different centralizer types, sizes and materials, spacings, wear rates and wear zones.

Centralizers and non-rotating centralizers stabilize the rod string, eliminate or reduce rod coupling wear and wear on tubing and reduce the torque in deviated well with greater rod and tubing service life. The results are energy savings, lower work over costs and less down time.

Centralizer manufacturers include: R&M energy systems, KUDU, Canam pipe & supply.

8.4 HOLLOW ROD DRIVE PCP

8.4.1 Tubing Drive PCP's

Small tubing can be used to drive the PCP's rotor but tubing coupling threads are not designed to transmit high torque to safely rotate the rotor. In the event of a sudden stop, the back spin takes a chance to easily unscrew tubing connections.

8.4.2 Hollow Drive Rod

The use of solid drive rods has become standard for PCPs, but new technological progress in terms of deeper wells producing greater quantities has meant an increase in early drive rod failures inherent to the limitations of the system. Premature failures are due to an increase in the working loads on solid drive rods as flexion-torsion combined stresses, overloading unions and tubing and rods friction wear.

New monolobe and multilobe high capacity PCPs (between 500-1,000 cubic meters per day and more) used today require a driving rod device that could transmit high working load and torque between 1,000/2,500 ft.lbf (135/340 daN.m) or more and cannot be safely driven with standard solid drive rods. (1 ft.lbf = 0.135 daN.m). See PCP High Capacities chapter 13-1.

The hollow drive rod is a NEW rod concept, designed to work under rotating loads with high torque capacity to increase pumping rate and enable injection through it.

Hollow drive rods are manufactured by Tennaris. The HolloRod™ Series combine a low-alloy Chromium-Molybdenum with a specific mechanical and heat treatment resulting in excellent mechanical properties, yield stress 140 kpsi minimum (964 MPa), ultimate tensile stress 147 kpsi (1,012 MPa) (1 kpsi = 6.89 MPa).

These hollow rods have box-box ends and a nipple connector with a specific thread that allows an external flush joint or small up set. The designed connector minimizes turbulence, flow losses and reduces friction between rods and tubing and premature wear failures. Specific hollow polished rods are used to drive the head.

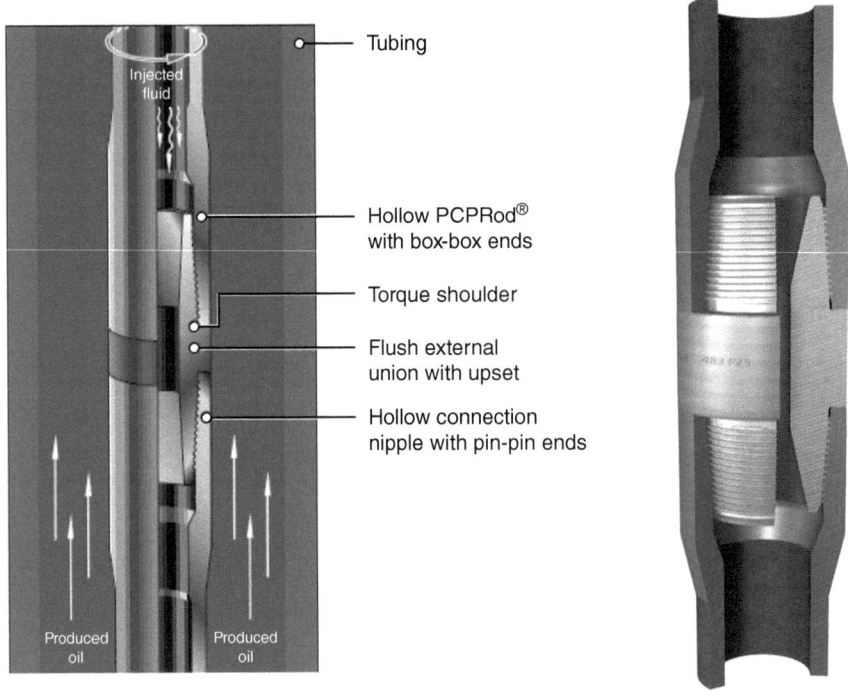

Figure 30

Hollow PCPRods™, progressive cavity pumping drive.
Source: *Tennaris*.

Tennaris PCPRod® technical specifications:

Tennaris PCPRod®	Outer diameter	Wall Thickness	Upset outer diameter	Joint min. int. diameter	Max. working torque*	Metric Weight
1000	48.8 mm 1.92"	6.7 mm 0,26"	No upset	20 mm 0.79"	1,000 ft.lb 135 daN.m	7.1 kg/m 4.77 lbs/ftr
1500	42.2 mm 1.66"	5.0 mm 0.2"	50 mm 1.97"	17 mm 0.67"	1,500 ft.lb 203 daN.m	4.7 kg/m 3.16 lbs/ft
2500	48.8 mm 1.92"	6.7 mm 0.26"	60 mm 2.36"	20 mm 0.79"	2,500 ft.lb 339 daN.m	7.2 kg/m 4.84 lbs/ft

(ft.lbf = 0.138 kg.m = 0.135 daN.m, 1 daN.m = 7.37 ft.lbf)

Hollow Drive Rod advantages:

− Work under high torque stress.
− Used with high volume PCP's.
− Increase pumping rates due to a high torque capacity.

- Less tubing wearing, due to the non-upset and minimal upset connectors.
- Possibility to inject different kinds of fluid *via* the inside of the "Hollow Rod". i.e. viscosity reduction fluids to pump extra heavy oil, chemical corrosion inhibitors, etc.
- Increase reliability and lower operating costs due to the elimination of connection failures.

Applications:

- High production rates (mature fields).
- Deviated wells.
- Heavy oil applications, diluent injection capabilities.
- SAGD, steam injection capabilities.
- Waterflood and others applications.

8.5 COILED TUBING DRIVE PCP

Coiled tubing (CT) is commonly used in well operation as sand cleanouts, stimulation, coil completions and drilling operations but coiled tubing can be used too as a drive string for PCPs: the CTPCP.

The coiled tubing drive PCP system is a NEW concept that integrates coiled tubing as a drive string instead of the classic solid rod string.

The coiled tubing has a dual purpose function. It is suited to transmitting torque and rotational movement to the PCP rotor and serves as a conduit for production up from the inside or down for injection of chemicals as diluents.

The most significant advantage of coiled tubing is the elimination of connections as the failure of connections in a conventional sucker rod string is the cause of work overs.

Adapted drive heads are used with coiled tubing.

The Coiled tubing to drive PCP pumps is an H.Cholet IFP Energies nouvelles patent "Volumetric Pump driven by a continuous tube", FR US 5,667,369 Sep. 16, 1997, CA 2 163 711 Apr 13, 2004.

Advantages:

- No coupling, reduced tubing wear.
- Possibility to inject different kinds of fluid by the coiled tubing as viscosity reduction diluents to pump extra heavy oil, chemical corrosion inhibitors, etc.

Coiled tubing OD sizes from manufacturers are: 3/4", 1", 1-1/4", 1-1/2", 1-3/4" 2", 2-3/8", 2-7/8".

8.6 CONTINUOUS DRIVE ROD

A continuous rod string is a round cross-section continuous rod to drive a progressing cavity pump.

Figure 31

Continuous drive rod and transport.
Source: *Weatherford*.

The manufacturer Weatherford provides continuous rods that are API Grade D carbon steel, chrome-molybdenum. Sizes and maximum lengths on a specific reel are 1" (25.4 mm)-2000 m, 7/8" (22.2 mm)-2600 m and 13/16" (20.6 mm)-3000 m. It only requires one bottom connection to the PCP's rotor and one at the top string polished rod, welded onsite during installation. It can be connected directly to the drive head eliminating the need for a polish rod.

The continuous rod is stored on an 18-ft (5.4 m) diameter reel, transported and installed at the well site. The rod is deployed into the well at a rate of 30 m/min by a variable-speed, hydraulically driven gripper mechanism. The continuous rods are more commonly used for pumping in USA and Canada.

Advantages: Continuous sucker rod compared to conventional threaded solid drive rod strings.

- No coupling.
- Reduced wear rate between tubing and continuous rod due to uniform load distribution.
- Larger annular space minimizes flow friction and pressure losses, decreases the power requirement.
- Lighter in weight.
- Fast retrieval, repair or replacement will minimize repair costs compared to drive rod strings.

Continuous rod strings are manufactured by Weatherford-EVI Oiltools – COROD™ and Oil Lift.

8.7 POLISHED ROD AND CLAMP

The polished rod (PR) is the uppermost joint in the rod string for connecting and transmitting the rotational movement from the PCP drive head to the drive rods string. It is specially designed for high tensional and torsional strength and recommended with high strength coupling.

The polished rod has an optimal surface finish to ensure an appropriate and efficient hydraulic seal to be made around the rotation rod in the stuffing box.

The rods are round metal core with a section diameter of 1-1/4" (31.8 mm) to 1-1/2" (38.1 mm) and currently made from different alloys: AISI 4130 and AISI 4140. API Polished Rod.

A 0.8 safety factor is recommended on drive rods, pony rods and polished rods to maximize fatigue life.

Polished Rod Block Clamps

The polished rod block clamp is located on top of the drive head to clamp the polished rod and secure it during pumping. The block clamp is tightened with two to four large bolts and sized with the polished rod used for a maximum working load.

Rod block clamps are available for polished rods sized 1" to 1-1/2".

Manufacturers include: Robbins & Myers Energy Systems, Harbisson-Fisher.

8.8 TUBING ANCHOR

8.8.1 Tubing Anchor Catcher

The tubing retrievable Anchor Catcher is located under the PCP's stator pump. It anchors the tubing string to the inner casing and holds it in tension to prevent movement of the tubing string during pumping operations.

Two types:

The mechanical-set tubing anchor catcher. It is a retrievable double grip tubing anchor designed to anchor the tubing string with the proper amount of tubing tension. A right hand tubing rotation releases the anchor. An emergency adjustable safety shear pin can release it by over pulling the tubing.

The hydraulic-set tubing anchor catcher: for example, Weatherford's MH hydraulic-set tubing anchor catcher is a retrievable catcher set by tubing pressure without tubing manipulation. The setting mechanism is a bidirectional slip anchor to prevent the tubing from moving upward and downward with a straight-pull shear release that enables release-force adjustment.

Figure 32-1

Tubing anchor catcher fixed to the PCP stator.
Source: *Halliburton*.

8.8.2 Tubing Torque Anchor

The "Torque Anchor" is a specific sub mounted under the PCP's stator to take the reactive torque given by the friction between the rotor rotation and the stator.

When the PCP rotor is running clockwise, the reactive torque applied by the stator to the above tubing tends to rotate and unscrew it.

To prevent the tubing from being unscrewed, a specific tool, the "Torque Anchor" is mounted under the PCP's stator (or above, depending on the model used) to prevent tubing back-off and to handle the increase in vibration when larger PCPs pumps are used. The "torque Anchor" is designed to be running on tubing and retrieval in a variety of downhole conditions including Deep Wells, Horizontal Wells and to produce Heavy Oil.

The torque anchor uses the principle of a mandrel with one or two level cams pushing out anchoring blades mounted in a case. Blades are permanently pushed and kept in touch with the casing wall by a mechanical device to block the clockwise torque rotation on tubing but it can rotate freely in an anticlockwise direction. The oil flows through the torque anchor central shaft to the PCP stator inlet.

The torque anchor is taken down in the well along with the PCP's stator. When it has reached the desired depth the torque anchor is set from surface, by rotating the tubing hanger clockwise. The clockwise rotation of about 300 ft.lbs of torque securely fit the anchor blocks in the casing wall. Torque can be managed and the pumping started. To disengage the torque anchor, just stop pumping and relax the initial torque position by counter clockwise rotation.

Remarks: using the torque anchor may scratch the inside casing and cause tubing deformation when it gets longer because of variation in reservoir temperature and the load change during production start-up.

Different torque anchors are available from size 4-1/2" to 10-3/4", depending on casing specifications, API connection and weight.

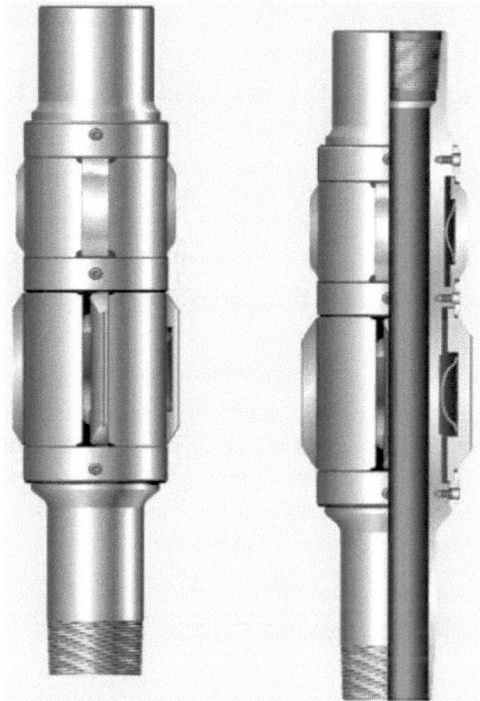

Figure 32-2

Centralizing dynamic torque anchor located under PCP stator.
Source: *Evolution Oils Tools Inc.*

Torque anchor manufacturers include KUDU XF Anchor-Cam-Loc Anchor, Canam Twister, Protex-CIS Centralizing Dynamic Torque Anchor®, PCP Oil Tools Dynamic Torque Anchor, Evolution Oil Tools Inc, Premium ALS, Frank Henry Equipment.

8.9 OTHER DOWNHOLE COMPONENTS

8.9.1 Remote Latching Between Rotor and Rods

The downhole remote latching device connects the drive rod string to the PCP's rotor. It is a male and female connector, with the male end attached on top of the pump rotor and the female end attached to the end of the rod string. The female is engaged and disengaged from the surface with a simple push/pull operation.

The remote latching tool is useful for removing and replacing the rod string if the PCP's rotor is too big to pass through the production tubing.

Advantage Products Inc provides the remote latching ON/OFF Tool™.

8.9.2 Tag Bar-Stop Pin

The "Tag bar" sub or stop pin sub is located between the stator end and the anchor sub.

The Tag bar stops the rotor end and drive string. The drive rod is lifted up to the correct rotor position inside the stator.

An incorrect rotor position can generate problems such as:
- The rotor installed being too deep. It runs in compression and binds on the tag bar. The rotor friction rotation with the tag bar generates heat damage to the stator lower elastomer, rotor breakage or a stator or tubing becoming unscrewed.
- The rotor is positioned too high in the stator. Pump efficiency is reduced, vibration is induced, the upper stator is damaged and the rotor suffers fatigue.

8.9.3 Rotor/Stator Position by Downhole Sensors

The Zenith PCP Protection Systems use downhole sensors with data transmission to the surface to determine the optimum position of the rotor inside the stator elastomer.

The system provides real-time information on the rotor position whilst the PCP is running. The real time reading monitors whether the rotor depth has changed due to increased load, thermal expansion or drive rod wind up or twist as changing torque and load is applied by the surface drive. The system also provides a real time downhole rotor RPM compared to the surface drive RPM speed.

8.9.4 Tag Bar Screen

The Tag Bar Screen from Tanroc is an "Ultra-Flow" screen at the pump inlet. It is adjustable from zero flow to 100% maximum perforation flow capacity that vibrates to prevent plugging and scaling.

8.9.5 Spring Loaded Tag Bar

The Spring Loaded Tag Bar from SCOPE acts as a slotted tag bar enabling fluid movement through the bottom of the tag bar to remove sand bridges at the pump intake.

8.9.6 No Go Tag on Top of the Stator

The Top Tag™ from KUDU is a sub located on top of the stator to stop the up end of the rotor on a NO GO seat. It allows the rotor position inside the stator to be easily located without using a bottom tag bar.

Same principle for rotor placement with the NO-GO TAG™ from National Oilwell Varco.

Advantages are:
- The rotor is never run in compression on the tag bar; no heat damages the stator lower elastomer.
- Leaves pump intake free of obstruction allowing the rotor to agitate solids.

8.9.7 Multi-Intake Sub

The Multi-Intake sub from National Oilwell Varco, NOVmonoflow allows the PCP to have more than one intake. More intakes give more inflow area; the flow of fluid is less restricted into the pump inlet for better performance and pump life.

8.9.8 Safety Valve and PCP's Pumps

In subsea offshore well production there is a strict requirement for a failsafe safety valve system to be employed. The safety valve can be closed automatically in case there is a problem.

The PCP system requires a rod string to drive the rotor. Standard safety valves such as flapper valves are unsuitable for PCP where the rod string is required to pass through the bore of the system down to the PCP rotor and typically the pump is located below what would normally be the location of a safety valve.

Caledyne Torus has developed a fail-safe closed Insert Safety Valve, which offers a permanent conduit through the centre of the valve. The valve utilises a sliding sleeve mechanism to shut in production operated *via* a power spring and piston arrangement. The rod drive rotation motion is transferred by rotational seals through the valve. This allows the PCP to be used with this safety valve. The valve is installed in the safety valve nipple profile and operated by the existing hydraulic control line.

Caledyne Torus available safety valve sizes are: 4-1/2", 5-1/2" and 7".

PCP and insert PCP pumps can be installed in wells requiring safety valves particularly offshore.

CHAPTER 9

Drive Head

9.1 DRIVE HEAD PRINCIPLE

On surface, the "Drive Head" electric or hydraulic motor powers in rotation the downhole PCP.

The drive head is connected to the drive string of rods which is interdependent of the pump rotor. The stator is screwed on at the production tubing extremity and generally held by a hanger at the wellhead. A tee joint is often fitted between the wellhead and the drive head (see Fig. 11).

The drive heads are designed in accordance with the load they must support and their fixing systems on the drive string and on the motors.

9.1.1 Drive Head Function

The drive string is supported by a drive head which is installed on the wellhead.

Its functions are:
- To transmit the rotational power movement of the motor to the drive string and the pump rotor.
- To seal the drive system from the well fluid on the tubing head and polished rod, by means of a stuffing box.
- To carry the axial load of the rods, and the head rating of the pump.
- To have sufficient braking energy to avoid backspin of the rod string in case the drive motor stops suddenly.

9.1.2 Drive Head Types

PCP manufactures and distributors offer three types of drive heads:

1. With a **one-piece shaft**: recommended in case of direct line coupling with a speed reducer (Fig. 33).

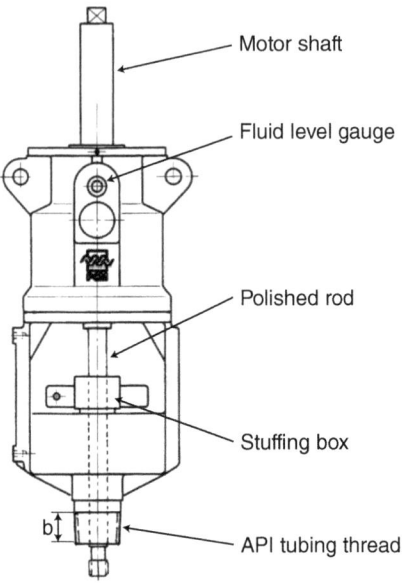

Figure 33

Drive head with solid shaft.
Source: *PCM*.

2. With a **hollow shaft** allowing a polished rod crossing screwed at the drive string extremity. This type of drive head is recommended for drive systems designed with belts and pulleys fitted on a vertical rotation axis.

3. Also with a hollow shaft, but with a return angle allowing the use of a drive system with horizontal rotation axis. (Fig. 34).

For each type, distributors offer various capacities of load on thrust bearing.

9.1.3 Rod String Twist and Back Spin Energy

The solid rod string has one end fixed to the PCP's rotor and the other hand fixed to the surface drive head. The rod under rotation increases the torque until the rotor pump starts to rotate and pump the fluid.

The solid rod considered as a circular bar of constant section is under tension and rotates under torsion deformation with a momentum force or torque (N-m) applied by the drive head. The rod is under torsion deformation, characterized by the angle of twist. We assume the angle of twist will vary linearly compared with the rod length from the drive head to the end the rotor.

Before the rotor starts to rotate, a certain amount of potential rod torsion energy is stored by the twisted rod drive. This energy will be released when the surface drive head is stopped.

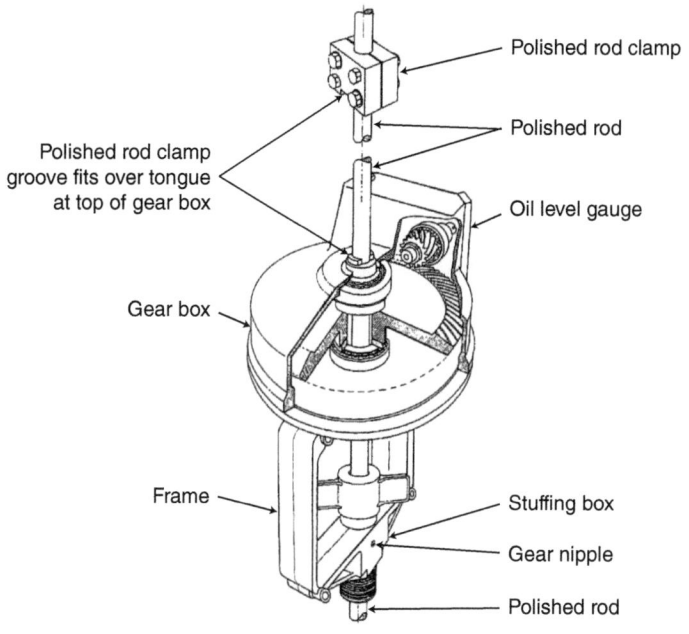

Figure 34

Drive head with hollow shaft and return angle drive.
Source: *Griffin Pumps*.

Many questions will arise:

- How much twist occurs in a rod string?
- When does the rotor start to turn at start up?
- What happens to rod twist when rod speed is changed?
- What happens with the stored energy in the rod?
- What are the rods' string rupture limits induced by the twisting and bending moment?

9.1.3.1 How Much Twist Occurs in a Rod String and when Does the Rotor Start to Turn at Start Up?

The number of rod twist turns before the pump rotor starts to turn depends on many parameters as rod grade and size, rod length, well deviation and rod friction to transmit the torque from surface, the end admissible torque being 75%.

Zenith Oil Field Tech provides a downhole PCP Protection system. It measures the number of turns of the pump rotor and on the surface in order to monitor the total number of twists in the entire rod string, to prevent back spin and rod breakage.

The rod twists change with the torque: the number required depends on many factors, including the speed in rpm, type of fluid, sand, temperature, pump characteristics and rotor/stator fit.

For example, the measured rod twist is 100 turns/1000 meters before the pump rotor starts.

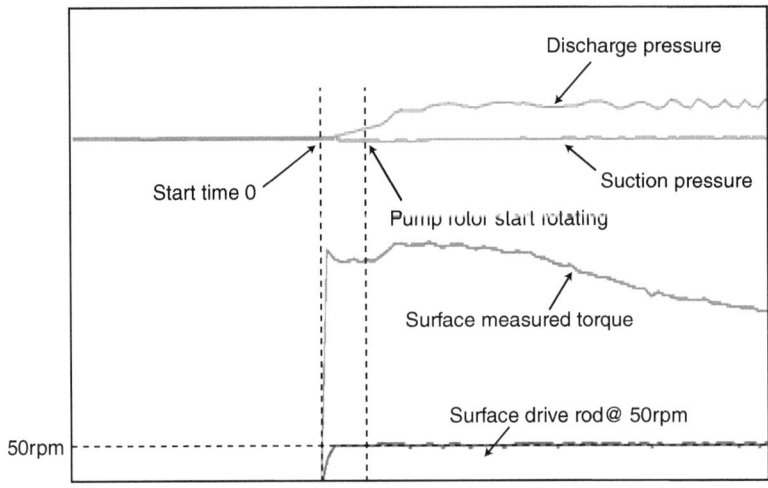

Figure 35-1

Rod twist, 100 turns on the surface before the rotor pump starts.
Source: *Zenith Oilfield tech Pump Protection system*.

9.1.3.2 What Happens with the Stored Energy or Back Spin Energy from the Rod?

When the drive head suddenly stops, the mechanical torsion energy stored in the rods string is released and drives the rods string and all moving parts in the drive head in the reverse rotational direction (100 turns for example). That is the "mechanical backspin".

If fluids are present inside the production tubing with a differential pressure across the PCP pump stator, the fluid potential energy stored is released through the PCP pump. The hydraulic PCP rotor torque activates the rods' string in the reverse rotation direction. That is the "hydraulic backspin".

By reaching high speeds two these back-spins could generate severe damage to the equipment and may cause accidents.

The backspin effect is controlled by a mechanically or hydraulically controlled brake system engaged automatically when the rod string begins to turn in the reverse rotation direction (left hand).

9.1.3.3 Evaluation of the Rotational Kinetic Energy Stored by the Twisted Rod Drive

We assume a steady state just before the rotor starts to pump with a 2000 N-m constant rod torque applied on the surface, 1000 meter rod length and 100 turns rod rotation (1 turn = 2×3.14 radians per 10 meter rod length).

- Per 10 meters of rod and one turn (2×3.14 radians), the rotational potential energy stored from rod deformation is $= 1/2 \times 2000 \times 2 \times 3.14 = 6.28$ kJoule.
- Per 1000 meters and 100 turns, the rotational kinetic energy stored $= 628$ kJoule $= 150$ Kcal.

If the back spin 100 rod turns is released for 60 seconds, the equivalent power to dissipate will be 10.46 kJ/sec = 10.4 kW.

9.1.4 Drive Head Brake and Safety Protection if Sudden Stop

A brake system installed in the surface drive head is: a mechanical friction brake (Fig. 35-2), a hydraulic brake by pumping liquid through the orifice resulting in a pressure drop or an electric motor brake. The brake and a controlled process release the mechanical torsion energy stored in the rods' string and the hydraulic fluid well potential energy. The brake system is used to slow down and stop the rods' rotation when the well is shut down as soon as the motor drive head is stopped. The braking force decreases as the torque in the well is released, allowing the fluid to drain from the tubing as completely as possible. The drive head brake system has to be safe and reliable with smooth backspin control eliminating wear on the brake components.

The drive head brake is rated to safely release the back spin energy changed in calories resulting in an increase in the temperature of the brake components, but not excessively if there is a risk of an explosive atmosphere in the pump environment.

Drive head PCP' manufacturers will comply with standards for specifications. The specifications should be clear and understandable for the pump operators who should be able to choose a safety drive head for its application.

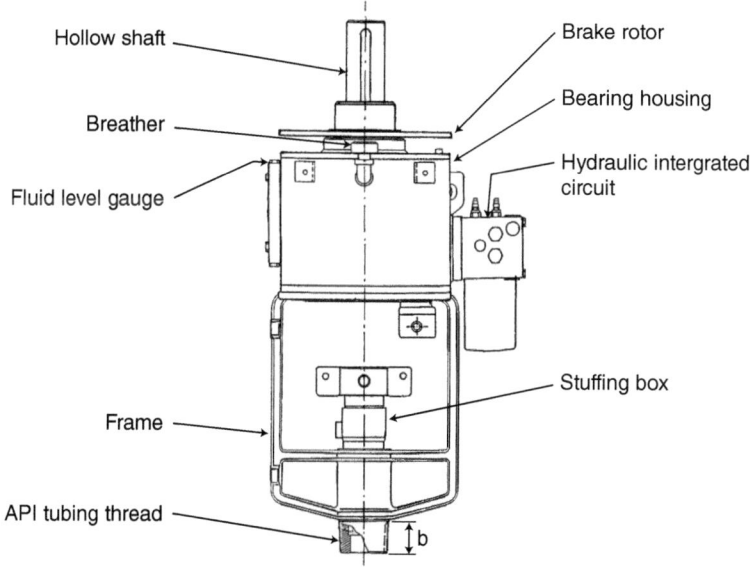

Figure 35-2

Drive head with back-spin brake rotor.
Source: *Kudu*.

Some rules are already in place:

1. The manufacturer should design the equipment with a 150% safety factor as regards the optimum working conditions, and should take into account the additional loads generated by impacts, as well as the effects of heat on the component's properties.

2. The manufacturer should supply a handbook which defines the maintenance operations for the braking system including the fluids replacement.

3. The operator should set a back-pressure valve at the wellhead so as to avoid any fluid back spin.

4. The operator should select a drive head holding a braking system with a higher maximum torque than the drive string resistive torque.

If the drive system is not installed vertically (as is the case for deviated wells), the bending moment generated by the inclination may increase some constraints which should be taken into account by the operator.

9.2 DRIVE HEAD SYSTEM

All types of drive systems may be used:
- Electric motor drives located at the drive head and coupled to the rods string.
- Hydraulic motor drives.
- Internal-combustion engines, gas or diesel driven.

These drive speed systems may provide fixed speed or variable permits for controlling speed and torque applied on the rod.

Some examples are shown on Figure 36.

9.2.1 Hydraulic Power of the Pump

The power of the pump P depends on the total resistant torque (mechanical + viscosity). The power with the mechanical torque is:

$$P(\text{watt}) = Torque(\text{N.m}) \times 2\pi \times N(\text{rotation/sec})$$

or

$$P(\text{kW}) = 1.046 \times 10^{-3} \times \Gamma(\text{daN.m}) \times N(\text{rpm})$$

where:
- P Hydraulic power (in kW)
- Γ Total torque (in daN-m)
- N Rotary speed (in rpm, rotation per minute)

or:

$$P(\text{HP}) = \frac{\Gamma(\text{ft.lbs}) \times N(\text{rpm})}{5252}$$

$P(\text{HP})$ Horsepower (1 Hp = 0.736 kW)
T Torque in ft-lbs
N Rotary speed (in rpm)

To protect the braking system from overload, the drive system equipment should be designed with a possible disconnection when it reaches the maximum capacity of the braking system.

It is necessary to measure the energetic capacity of the drive head braking system. So, the operator should select a braking system with a higher energetic capacity than the well-head drive systems.

Manufacturers offer a selection of braking systems that can be used on the drive heads:

– Centrifugal braking systems.
– Automatic brake with adjustable disk.
– Mechanical brake manually controlled.
– Hydraulic brake.

Also, it is possible to use a torque limiter so as to disengage the motor if the torque reaches a pre-determined value.

Drive Head Examples

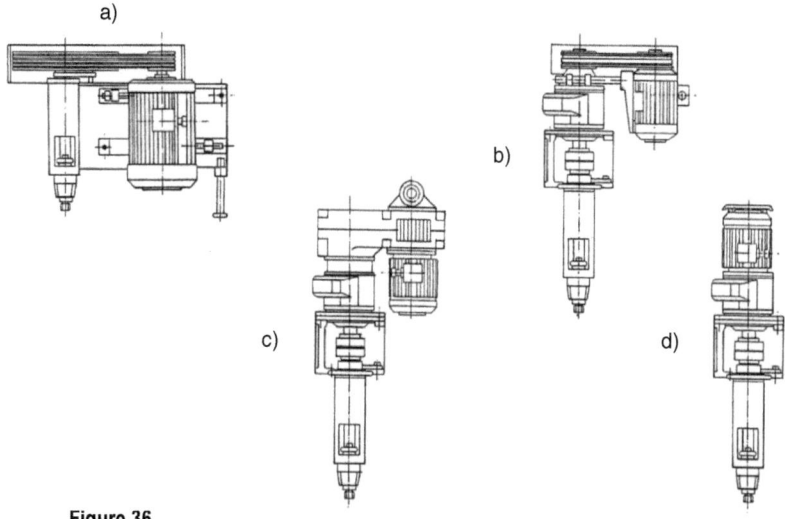

Figure 36

Examples of drive systems. Fixed speed drive: belt drive (**a**), belt and gear reducer drive (**b**), Variable speed drive: mechanical system (**c**), variable frequency direct drive system (**d**).
Source: *PCM*.

9.2.2 Direct Belt Drive Head

The mechanical belt drive's (Fig. 36a) advantages are simplicity and low cost. But it should be used only for medium and high speeds. To change the speed, the motor has to be stopped to change the belt on the sheave's reduction ratio.

For example, assuming a reduction ratio of 5 to 1, the minimum speeds are:

	Frequency	
	50 Hz	60 Hz
4 pole motor	300 rpm	360 rpm
6 pole motor	200 rpm	240 rpm

9.2.3 Mechanical Belt and Gear Reducer Drive Head System

The combination of a belt drive system with a gear reducer (Fig. 36b) allows lower speeds to be used.

9.2.4 Variable Speed Drive Head (VSD)

The use of a Variable Speed Drive allows the production of the well to be optimized.

Variable Speed Drive Advantages

PCP is a "constant flow machine". The pumped flow rate is proportional to the rotating speed (RPM) and the reservoir conditions require that the flow rate be frequently varied in response to the dynamic's fluid level; the same applies to sand slugs increasing friction and torque to turn the PCP rotor.

Variable speed rod drives and controlled torque applied on rods are essential in order to adapt the pump rate to the well production. The variable speed drive means that a rotational speed can be chosen that is compatible with a minimum submergence level, hence a maximum flow rate. Unstable well production can be safely carried out using PCP with VSD particularly when pressure drops are difficult to accurately determine. In such cases, it is possible to start and stop the pump and to adapt the rotating speed in keeping with the pump's production rate.

VSD optimized the pumping system by permanently adjusting interdependent parameters and criteria such as motor speed and torque, power demand, flow line pressure.

Pumping system results are extended pump run life and better pump efficiency with substantial operating and capital cost saving.

Different types of variable speed drive are available.

9.2.5 Mechanical Belt Drive System

The variable speed drive mechanical system consists of two variable cone sheave diameters on which a belt can move and thus enable speed variation of the shaft linked to the drive rods. Speed change ratios from 1 to 6 can be generated.

9.2.6 Hydraulic Drive System

A hydraulic power unit pump driven by an electric or internal-combustion engine, gas or diesel generates hydraulic power fluid energy to operate with a hydraulic motor to turn the polished rod and the pump.

The drive string should be protected from excessive torque by means of a hydraulic pressure controller and safety valve.

Advantages are no belt and sheaves or gear, a wide range of speeds are available.

But the hydraulic drive intermediate power requirements might be more costly.

9.2.7 Direct Drive Head and Permanent Magnet Motor

The specificity of a Direct Drive Head is the head geometry. The motor is positioned vertically with concentric axis, in-line with a polish rod and connected through a stuffing box to the rod string. Compared to standard head belt or gear drive, there is no gear or mechanical part of belt transmission for reduction speed to drive the rod string at low speed. The results are fewer components with fewer moving external parts, such as belts, drives and sheaves to act as a torque multiplier and/or speed reducer and easy maintenance because no belts.

9.2.7.1 NEW: The Brushless Permanent Magnet Motor to Direct Drive Rods

Brushless Permanent Magnet Motors (PMM) are increasingly used in industrial equipment such as transport and electric cars.

Permanent magnet electric motors with direct drive head have been recently used instead of standard asynchronous induction electric motor.

Permanent magnet motor design is fundamentally different from the standard AC asynchronous induction motor. The PMM is a synchronous motor, three-phase winded stator and rotor poles with multi-permanents magnets. PMM can deliver 10%-30% improvements in overall system efficiency over AC induction motors. The increased motor efficiency is partially due to not having to generate a stator magnetic field, which is instead generated by the permanent magnets.

These PMM direct drive heads are vertical hollow axles directly connected to polish rods through a stuffing box to the directly driven PCP rod string.

The PMM is powered by a specific Variable Speed Drive (VSD) frequency controller to provide high torque at the desired rod speed range integrating control algorithms to increase production, improve energy efficiency and enhance the reliability of artificial lift systems.

9.2.7.2 Direct Drive Head Benefits

Benefits in driving the PCP with a direct drive head with permanent magnet motors are the use of recent brushless magnet synchronous motor technologies instead of standard asynchronous induction electric motors.

- Direct drive vertical hollow shaft, concentric with the rod string allowing a polished rod crossing the direct drive head.
- Permanent magnet motors are controlled by electronic variable speed drive (VSP).
  - Benefits are large variable speed range adjustment examples from 35 rpm to 450 rpm and high controlled torque at low speed,
  - A maximum torque control capabilities with smooth start and shut down of speed motor.
- To start the pump the VSD power controller gradually adjusts the power current compared to the high inrush current needed to start the typical induction motor.
- Precise torque and speed control can avoid broken rod and trip issues.
- Higher efficiency compared to induction motors.
- Electrical power energy saving is about 15%, consumption is reduced.
- The magnet motor means that torque can be controlled to prevent backspin. After the motor stops, the drive rod safely releases the torque to counter-rotate the rotor. Under the electric regenerative braking system, the stator coil generates an electric current that rotates the rotor and balances the counter-rotation torque of the drive rod. The electric regenerative braking system is dissipated in bank resistor.
- Safe during pumping and braking operation, the moving parts are internal.
- Low noise level, less maintenance.
- Lighter than standard drive head.
- Reduced maintenance downtime.
- Well production optimized with PCP flow rates regulated by a wide range of variable speeds.

These NEW types of surface direct drive head with magnet motors are recent, developing technologies on a promising market.

Permanent magnet motor technologies are also used downhole to drive the ES-PCP see chapter 12-3.

9.2.8 The Electronic Variable Speed Drive (VSD)

The variable speed is provided by a frequency converter. The usual frequency range extends from 10 to 100 Hz, thereby providing a 1 to 10 speed range ratio. A speed reducer installed between the motor and the drive head enables the speed range to be adjusted.

The speed can be changed whether the system is running or not.

The advantage of this system is that it may be remotely controlled by a downhole pressure sensor allowing the rotating speed to be adapted to the submergence level.

Electronic Variable Speed Drives are compatible with asynchronous and permanent magnet motor.

9.3 DRIVE HEAD CODE ISO NORM

The ISO 15136-2:2006 standard provides requirements and information on progressing cavity pump surface-drive systems. Informative annex B provides information on the brake back spin system, installation, and operation, as well as drive rod selection and use.

9.4 DRIVE HEAD FROM MANUFACTURERS

The driving head proposed by many manufacturers presents many technical advantages which are summarised below:
- All drive heads are standard connections for vertical well and for deviated or slant wells.
- Drive head power transmissions are belt or gear drives to a bearing box and hollow-shaft with internal backspin control when the drive head stops.
- Safety back spin brake type centrifugal or others are mechanically activated to control back spin under any conditions.
- A removable stuffing box maintains the well head pressure.
- Economic advantages include reduced capital, less set-up time, simple installation and maintenance, reduced expenses and power efficient, smaller footprint and increased reliability.

9.4.1 Belt or Gear Electric Standard Drive Head

Figure 37-1

Belt drive head.
Source: *Oil Lift Technology*.

Figure 37-2
Drive head for slant well.
Source: *Kudu*.

Figure 37-3
Electric and hydraulic drive head.
Source: *GrenCo*.

Many Others Manufacturers

CAN-K, Baker Hughes Centrilift drive head LIFTEQ PCP. INDRIVE

9.4.2 Gear Drive Head with Hydraulic or Electric Motors

A gear box reduces speed instead of a system involving pulleys and belts. The PCP flow rate is adjusted to the dynamic fluid level, and the torque to the friction by the rod speed. The variable rod speed is obtained by the motor speed, hydraulic type or electric powered with a variable frequency supply.

9.4.2.1 Gear Drive with Hydraulic Motor

The hydraulic power unit provides high pressure hydraulic fluid to the hydraulic motor. The hydraulic motor is positioned vertically and bolted to the speed reduction gear box, driving the rod with variable speed.

Advantages are the use of standard hydraulic motors in the range 70 hp at 1,200 rpm input speed and a gear box Figures 37-4, 37-5, 37-6.

Figure 37-4

Hydraulic drive head, VHGH 60HP.
Source: *Kudu*.

Figure 37-5

Hydraulic vertical drive head.
Source: *Weatherford*.

Figure 37-6

Moyno hydraulic Gear Drivehead (MGD™).
Source: *R&M Energy Systems*.

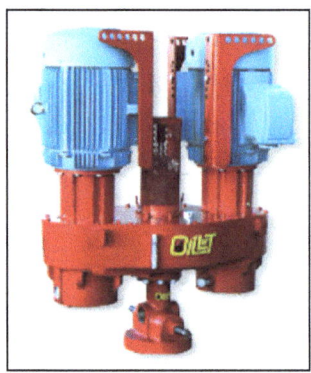

Figure 37-7

Gear drive.
Source: *Oil Lift Technology*.

9.4.2.2 Gear Drive with Electric Asynchronous Motor

A standard asynchronous induction electric motor (one or two) positioned vertically and bolted to the reduction gear box. The motor is powered by a specific variable frequency power supply.

With 1750 rpm, three-phase 460 VAC and 10:1 gear reduction ratio motors, the output shaft speed will be 175 rpm Figure 37-7.

9.4.3 Direct Drive Head with Permanent Magnet Motor

Direct drive head manufacturers

- **Daqing Jinghong Petroleum Equipment Factory** (China)

Specifications:

Motor sealed explosive proof Power	5.5 to 55 kW
Rated Load	150 kN
Polished rod diameter	22 to 42 mm
Rotary speed	0 to 400 rpm
Type of speed adjustment	Variable frequency
Pressure capability of well head	0 to 6 Mpa (mechanical seal)
Anti-counter rotating device	Electric-magnetic brake
Well head connection	7-1/16" flange

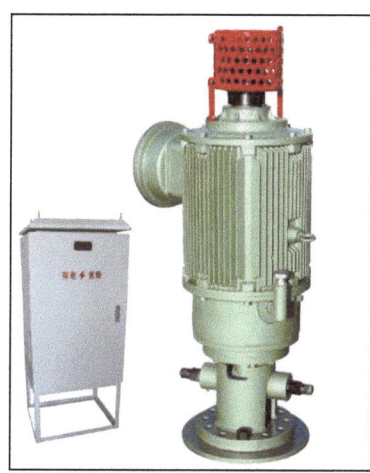

Figure 38-1

Direct Drive PCP head with permanent magnet motor.
Source: *Daqing Jinghong Petroleum Equipment Factory*.

- **Shengli Electric limited Company** (China)

Specifications:

Motor sealed explosive proof Power	15, 20, 30, 40 HP
Voltage	AC 380-660-1140 V/50 Hz AC 460-690/60 Hz
Polished rod diameter	44 mm (1-3/4")
Rotary speed	0 to 400 rpm
Type of speed adjustment	Variable frequency
Temperature	– 30°C – + 50°C
Noise	65 dB

Figure 38-2

Direct Drive PCP head with permanent magnet motor.
Source: *Shengli Electric limited Company*.

- **Advantage Products Inc** from Canada

Figure 38-3

The TorqDrive™. Direct drive head PCP with permanent magnet motor.
Source: *Advantages*.

The TorqDrive™ 450/1000 PC is powered by a variable frequency three-phase 460 VAC and 100 amperes, speed range 30 to 450 rpm, weight 600 kg.

9.4.4 Drive Head PCP's Manufacturers Model Identification

The drive head PCP's identifications are different from those of manufacturers as shown below:

PCM/Kudu

- VH 200 HP 18T: vertical, hollow shaft, electric 2 × 100 hp dual motors, 18 tonnes axial load capacity.
- VHGH 60 HP: hydraulic drive 60 hp.

Baker Hughes Lifteq

- LT-100E: electric 100 hp motor, 600 max rpm, max torque rating 1,932 N-m (1,425 ft-lbs), thrust bearing rate 50,300 lbf.
- H-1800: hydraulic motor and gear, 450 max rpm, max torque 2,712 N-m (2,000 ft-lbs).

Grenco

- D-2100: direct well head drive, max torque rating 2,845 N-m (2,100 ft-lbs), trust bearing rating 25,900 lbf.

Protex

- PDH-100: electric 100 hp motor, 800 max rpm, max torque 3,245 N-m (2,400 ft-lbs).

Grenco MonoFlow

- D-2100: electric with 100 hp single motor or 150 hp dual motor, 600 rpm max, torque rating 2845 N-m, (2,100 ft-lbs).

Netzsch

- DH (direct drive belt), GH belt = gear reducer drive).
- RH (belt + gear right angle drive).

PCP OilTools

- RRH-75HP-9T: electric 75 hp motor, 9 tonnes axial load capacity.

Indrive

- Inline beltless solid shaft direct drive.

Surface Direct Drive Permanent Magnet Motor

Advantage Products Inc

- TorqDrive™ 450/1000 PC. 1,350 N-m from 30 to 450 rpm.

Daqing Jinghong Petroleum Equipment Factory (China).

Shengli Electric limited Company (China).

9.5 ROTARY SEAL STUFFING BOX

The progressing cavity pump drive heads require a rotary seal stuffing box to make the seal between the polished rotating rod drive and the well head passage. This is so as to maintain the well head pressure and to prevent oil from leaking onto the ground.

The rotary seal's main constraints are the fluid to be sealed and the polished rod rotation and misalignment.

- The crude oil produced is light to heavy with a high water cut and typically contains fine sand particles. In others cases, fluids can include sand laden oil, SAGD thermal HT oil, dewatering gas wells.
- The polished rod misalignment. The PC drive heads run continuously and the alignment between the rotating polished rod and the rotary seal is imperfect.

The rotary seal can be fitted to most of the top drives available in the market.

Rotary seal manufactures include:

Weatherford:

The DuraSeal-500 rotating stuffing box with an innovative seal design, a redundant sealing and robust bearing system enables lower shaft run-out. The maximum static pressure rated is 20 bar (3000 psi).

Kudu:

The HT Zero-Leak Rotary Seal Oryx: It is low maintenance, zero leak tolerance for a minimal environmental impact. Suitable and proven for high temperature in thermal well SAGD applications, tested to 315°C (600°F) dynamic.

CAN-K:

The Zero-Leak stuffing boxes features, high temperature capability of up to 250 deg Celsius, dynamic pressure – 1000 psi, static pressure – Max. 3000 psi, handles high gas, can be custom designed to adapt to H_2S and other sandy applications, max. rpm of 1000.

Figure 39

Rotating seal Stuffing Box.
DuraSeal-500 Rotating Stuffing Box. Zero leak stuffing Box.
Source: *Weatherford*. Source: *CAN-K*.

9.6 ROD LOCK BOP

Part of a wellhead drive head, a specific BOP with lock ram is located between the tubing head and the bottom of the PCP drive head. To service the stuffing box or drive head, the BOP rams are closed onto the polish rod to prevent it from moving and to leave the well head isolated and secured. The stuffing box and drive head can be removed without the help of a winch line to hold the polished rod.

Advantages are: safer, faster, reduced rig time and cost savings compared to traditional methods.

Figure 40

Rod Lock© BOP.
Source: *OilLift Technology*.

9.7 TUBING ROTATOR

The tubing rotator is between the well head and the BOP. A specific tubing hanger located inside the tubing rotator replaces the need for a conventional tubing hanger.

While pumping with PCPs rod string rotating inside the tubing, the "Tubing Rotator" continuously rotates right hand the tubing to provide an optimum wear distribution around the entire internal circumference of the production tubing. The tubing rotation operation can be driven manually, electrically, hydraulically or with the drive head assistance. The number of rotations per day is dependent on the type of drive system installed.

Figure 41

Rodec™ Ultimate Tubing Rotator Spool.
Source: *Rodec/R&M*.

The tubing rotator is useful for horizontal, directional and slanted well completions, as well as older conventional completions. The uniform wear distribution inside the tubing will increase the tubing life span and will reduce operating costs for well service jobs and mean less lost production.

The Tubing Rotator Rodec™ is manufactured by R&M Energy Systems.

CHAPTER 10

Installation, Operation and Maintenance

This chapter outlines the recommendations for the installation and/or replacement of all major components which make up Progressing Cavity Pumps when the stator is fixed at the production tubing extremity, and when the rotor is driven from the surface by rods (Fig. 11).

10.1 INSTALLATION

The installation of Progressing Cavity Pumps is fast and easy. It is carried out by using standard oil field equipment.

10.1.1 General Consideration

The pump shall be installed below the dynamic level, because the pump requires a positive inlet pressure to operate efficiently. However, permanent lubrication of the stator is necessary to avoid any elastomer failure. Recording the annular level is therefore recommended.

In situations in which there is a high GOR (gas/oil ratio), running down the pump below the estimated bubble point level, or the nearest above it is recommended. It will then achieve better oil production performance, and better pump lubrication.

10.1.2 Pre-operational Checks

The well must be cleaned and cleared of any sand to 3 or 5 meters below the installation pump level.

However, make sure that all appropriate fittings and adapters are available at the well site in order to make the various connections:
- Stator to tubing.
- Tubing to drive head.
- Rotor to solid rod drive string or hollow rod.
- Drive string to drive shaft.

A number of coupling rods (box-by-pin) and of pony rods should be provided to compensate for the length difference between the tubing and the solids rods when installing the drive head. Pump suppliers specify the thread diameters of their equipment.

10.1.3 Running in Stator and Tubings

10.1.3.1 Checking

1. Take special care in the measurement of all the downhole parts:
   - Stator from the rotor stop, located at the lower end screwed onto the rotor base, to the upper end of the stator.
   - Tubings, as they are being fastened.
   - The fittings that are to be mounted between the tubings and the drive head.
2. Note all measurements.

10.1.3.2 Running in

1. Connect the rotor stop to the stator base.
2. Clean and dope all tubing threads, to prevent accidental unscrewing of the drive string. This is essential, especially when dealing with very heavy oil in cold weather.
3. Connect the stator to the first tubing.
4. Run in tubings down to the chosen depth.
5. Mount the hanger onto the wellhead.
6. Mount the tee for connecting the surface flow line.
7. Screw the adapter fitting on the drive head.

In some cases. it is possible to fasten on the stator base:
- Either a tubing with an anchor packer.
- A static gas separator.

10.1.4 Running in the Rotor and the Rods String

10.1.4.1 Number of Rods, Pony Rods, Sub Couplings

To determine the number of rods required, proceed as follows:

1. Divide the length of the tubings expressed in meters by 7.6 m or by 25 feet (length of a solid rod drive).
2. Retain the number of lengths of rods needed to constitute the drive string.
3. Add an extra rod to support the drive string, after running in and positioning of the rotor. The combined length of pony rods and sub couplings to compensate for the distance between the upper rod and the drive head.
4. For wells equipped with a hollow shaft drive head, adjust the total length with a polished rod.

Chapter 10 • Installation, Operation and Maintenance

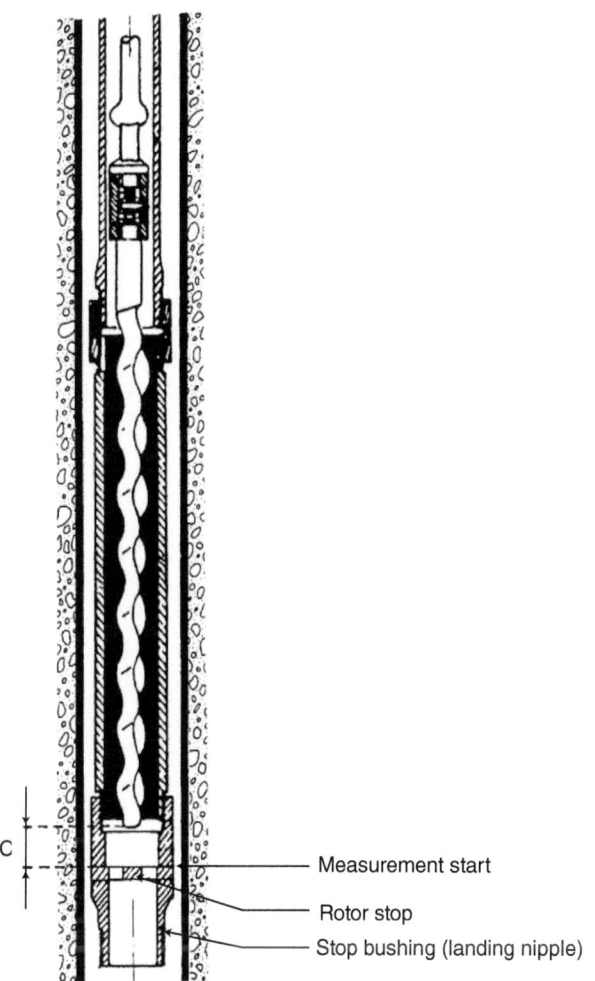

Figure 42

Running in the rotor/stator.
Source: *IFP Energies nouvelles*.

10.1.4.2 Running in the Rotor and the Rods

1. Connect the rotor to the first rod.

2. Run in all the drive strings by tightening to manufacturer connection specifications.

3. After the last rod has been screwed in, run in very slowly and watch for the drive string to rotate, indicating that the rotor has entered the stator.

4. Lower a further 1 m, then pull back up very slowly. The reverse rotation should then confirm that the rotor has entered the stator.

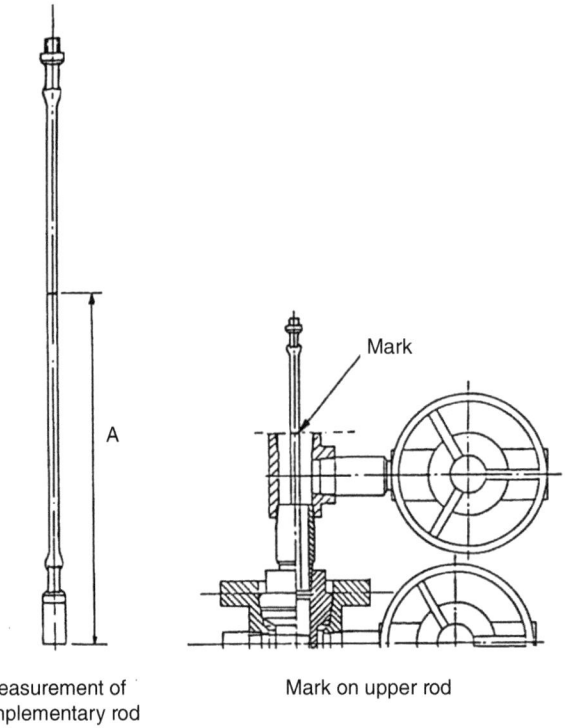

Figure 43

Running in rods.
Source: *IFP Energies nouvelles*.

5. Resume running in the rotor very slowly, and watch the hanging weight indicator so as to notice the time when the rotor lands on the lower end of the stator (indication of hanging weight decrease).

6. Lift the rotor very slowly so as to release it from its supporting point (hanging weight increase). Mark the rod at the tee level or at the fitting located below the drive head (Fig. 43). Repeat the operation to check.

7. Pull drive string up and remove upper extra rod.

8. Measure the distance A between the mark and the lower part of the sucker rod that has just been removed (Fig. 43).

10.1.4.3 Rotor Positioning Adjustment

The following values should be considered:

1. The extra tensile stress exerted on the drive strings by the pump head rating. This value is calculated from the *"load on thrust bearing"* relation mentioned in paragraph 3.1.5. The strain a is thereafter determined as a function of the rod section:

$$a = \frac{W \times L^2}{E \times S}$$

where:

- W weight per meter of drive string
- L length of drive string
- E Young Modulus
- S section of drive string.

According to the characteristics of their pumps, manufacturers have established a table for a direct reading of this value.

2. The distance b measured on the drive head between the lower end of the shaft (thread excluded) and the upper end of the drive head thread.

3. The height c by which the rotor base will be backed off of its supporting point (Fig. 42).

4. The distance d is the dilatation of drive string at the well temperature, when the tubing is anchored with a packer.

The new distance is: $B = A (a + b + c + d)$.

The pony rods and the couplings allowing this distance B should be selected.

Case of Hollow Shaft Drive Head

1. Insert the polished rod through the wellhead assembly until it extends about 30 cm from the bottom. Then put it in a vertical position following the previous calculations.

2. Set the clamp on the rod on top of the hollow shaft.

3. When using a new polished rod, wash the clamp area with solvent to remove the protective oils that may cause slippage.

10.1.5 Setting up of the Drive Head and the Motorized Driving System

10.1.5.1 Checking the Direction of Rotation of the Electric Drive System

The following operations should be performed:

1. Connect the power supply leads to the motor and switch on.

2. When the motor is running, check that the shaft rotates clockwise when looking down the well.

3. Mark connections and disconnect.

10.1.5.2 Setting

1. Attach a short pony rod on the top of the drive head.

2. Screw and tighten the upper coupling of the rod string to the drive head shaft that will thereafter be mounted on the wellhead.

3. Rotate the mounted assembly very slowly.
4. Connect tee to surface flow line.
5. Remove the pony rod and mount the chosen drive system.

Case of Hollow Shaft Drive Head

The polished rod shall be straight and free from corrosion, toll marks or other surface defects which might damage the stuffing box seals.

10.2 START-UP

10.2.1 Pre-start Check

1. Carefully connect the power supply leads by using the marks made previously.
2. Check the lubrication and the sealing of the bearings and stuffing box.
3. Open the flow line valve at the wellhead.
4. The PCP is ready to operate.

10.2.2 Operating Procedures

1. Switch on the motor. If the motor is provided with a variable speed drive, start at low speed, and increase speed progressively until the determined speed is reached. A time lapse corresponding to the filling of the tubing is required before the well fluid reaches the surface.

2. The rotating speed of the pump should be adjusted to the well productivity. Consequently, the dynamic or submergence level will be controlled frequently at the beginning of the operation.

3. Once the optimum speed is established, it will be maintained continuously and frequent on/off operations should be avoided.

If there is sand in the produced oil, it is preferable to have:

– A high rate of flow in the tubing.

– A large capacity pump with a low rotating speed (< 250 rpm).

10.3 INTERVENTION TIME

As an example, the starting-up of a pump at a depth of 1000 m m required the following operation times :
- Running in stator + tubing   2 h
- Running in rotor + solid rod   1 h 30 min

- Fastening of drive head   1 h 30 min
- Fastening motor + connection   1 h 15 min

Consequently, the installation and the starting-up of a PCP does not exceed one day of work-over operation. The experience that some operators have gained means that they are able to reduce these times.

10.4 OPERATING MAINTENANCE

The Progressing Cavity Pump system is simple, rugged and reliable. It was designed to require minimum maintenance, whatever the climatic conditions.

Adjustment and lubrication of the stuffing box should be carried out each month and a little less frequently for the bearings, by using products recommended by the manufacturer.

The dynamic level should be periodically checked to ensure optimum operation.

For PCP's failures and operating problems, see chapter 17.

CHAPTER 11

Insert PCP Systems

11.1 AN ECONOMICAL COMPLETION

Standard Progressing Cavity Pumps driven from surface by drive strings are designed very simply: the stator is fitted in the bottom of the production tubing, and the rotor is hung on drive strings (solid or hollow rods). The rotor rotation of the pump is generated by the drive head motor located at the surface and coupled with the drive strings.

The standard tubing run PCP concept requires two successive operations of lifting and running in for setting a new pump in the well during a work over. The first operation is the installation (or removal) of the rotor connected to the drive string and the second operation is the installation (or removal) of the tubing including the stator at its end.

But when a low production well is involved, the operator wishes to carry out the pump changing operation at the lowest cost. In order to substantially reduce the work over time and ensure good levels of profitability with a Progressing Cavity Pump, the "PCP insert pump or Insertable PCPs" concept has been designed.

The Insert pump system allows the PCP's stator-rotor pump to be installed (run in) and retrieved (run out) in a single operation without lifting the tubing by using the drive rod string. The "work over" operation is thus inexpensive and fast.

For small diameter pumps located at 1000 meters, the pump and drive strings assembly does not exceed 5 tonnes. In such conditions light lift operating equipment such as a rod service rig is sufficient.

Insert progressing cavity pump with latch and seal on the top stator from Kudu.

11.2 PCP INSERT COMPONENTS AND OPERATING PROCEDURE

Figures 44-1 and 44-2 represent a longitudinal section of this pump set at the production tubing end. As the tubing is lowered into the well, an anchor shoe is fitted which enables the seating of the proper pump. The tubing may be equipped with centralizers or anchored into the casing.

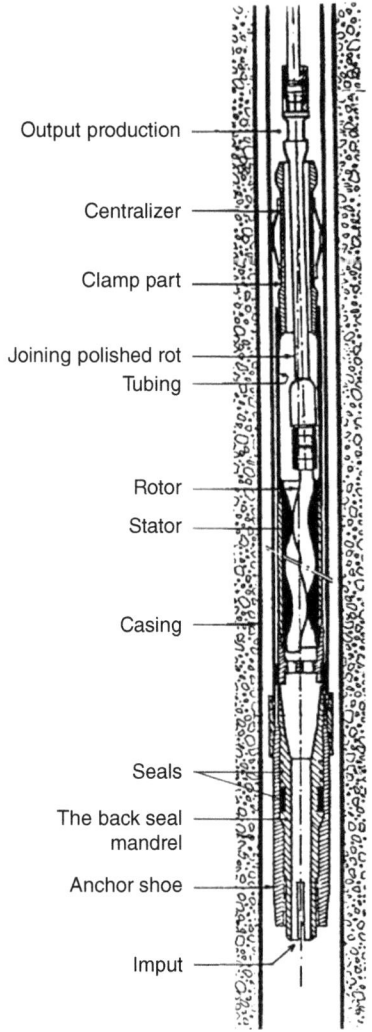

Figure 44-1

Basic principle Insert progressing cavity pump Latch and seal at the bottom of the stator.

Source: *IFP Energies nouvelles*.

The complete insert pump includes:

- A sub pump seating nipple – part of the production tubing – is positioned at pump depth.
- The stator pump insert assembly includes:
  - A vertical anchor locking device positioned on the top or at the bottom of the stator element. The anchor, (for example a J slot shoe type anchor auto locating top hold down or rotation torque anchor), automatically locates and securely seats the pump on depth in

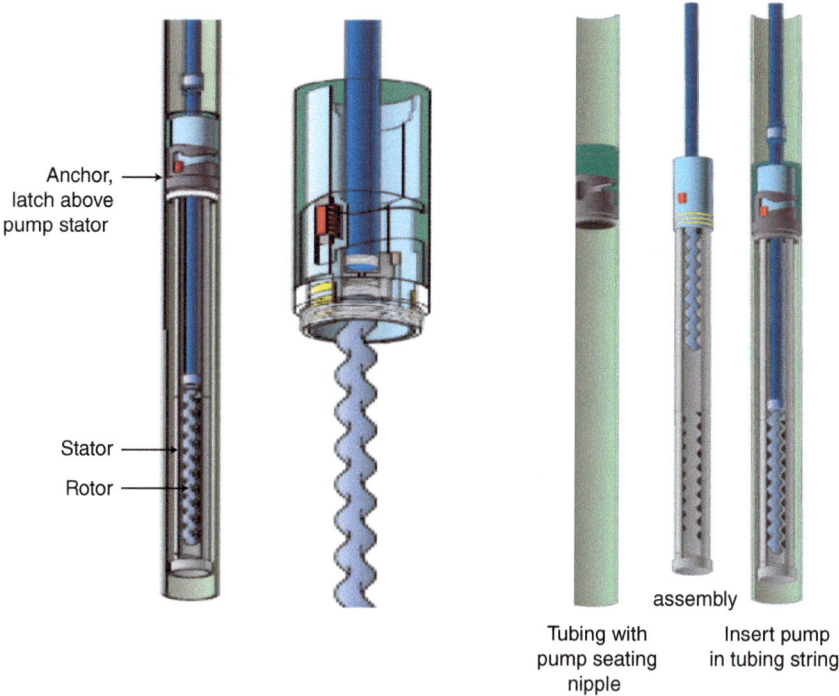

Figure 44-2

Insert PCP Systems.
Source: *Kudu*.

the pump seating nipple. The anchor device principle is a non-rotating mandrel to prevent axial and rotational movement of the pump assembly during pumping operation.
- A sealing assembly positioned at the bottom or at the top of the pump stator makes a seal between the stator and the inside tubing.
- The pump stator as standard.
- A specific coupling join is fixed between the rotor and the rod string to adjust the rotor position from the surface inside the stator and to set or unset the pump stator from the seating nipple.
- A by-pass sliding sleeve allows fluid flow through the assembly by pulling the rod string without unseating the pump. A flushable insert pump reduces the running out times in sandy and heavy oil applications.
- Parts of the insert pump are adapted to different pump sizes.

Operating Procedure

- Running in the tubing, the seating nipple is positioned at planned PCP stator depth.
- Both the PCP stator and rotor assembly are lowered down with the drive rod string, stator set anchor and seal on the seating nipple.

- To unset the pump insert from the seating nipple, just pull out the drive rod from the surface. The stator and rotor are retrieved with the drive rod to be changed.

11.3 PUMP INSERT ADVANTAGES

All the pumps offered by manufacturers are usable in "Insert" systems.

However, this pump should be run in tubing having an inner diameter that is greater than the pump stator diameter. Considering this, Pump Inserts are available from manufacturer for tubing size 3-1/2", 4-1/2", 5-1/2", but depending on tubing size and operating conditions, Insert pumps are mainly used for low producing wells which only require cheap equipment.

PCP insert pump advantages are:
- Flexibility:
  - The entire pump assembly can be installed and removed with the sucker rod string from the well, without tripping out the tubing when it needs to be replaced.
  - Pump positioned inside the production tubing and set on depth at the production tubing with a specific anchor shoe.
- Reducing overall life cycle cost.
  - Light operating equipment with a rod service rig is sufficient. There is no need for a service rig to pull out the production tubing to change a downhole pump.
  - "Work over" operation is not costly, reducing pulling costs by 40-50%, and fast, reducing downtime and rig expenses.
  - Faster replacement than conventional and reliable production.
- Minimizes the need to remove production tubing, reducing the cost of downhole gauge installations.
- Ability to use any progressing cavity pump without carrying out any modifications.

But:
- Sand may complicate operation of the pump assembly seal and latch.
- Pump handling can be a limitation. The reduced PCP pump size limits the pump capacity compared with a standard fixed end PCP stator on the tubing because a PCP stator insert has to move through the tubing.
- The well can be completed with a larger tubing for the use of a larger insert pump.
- The PCP insert pump has lower capacities than standard in same casings.

11.4 PUMP INSERT MANUFACTURERS

- Weatherford: Insert Pumps flushable Arrowhead™.
- KUDU: Insert Progressing Cavity Pumps Systems.
- National Oilwell Varco: The NOV Monoflo Insertable PC Pump System.
- R&M: Moyno Insertable Progressing Cavity (IPC™).
- The Premium Insert PC Pump Torque Anchor, dual anchor.

CHAPTER 12

Electrical Submersible PCP

12.1 HISTORIC

History of PCP Driven Downhole with Electric Motors

In the seventies, the PCP Moineau pumping principle driven by electrical submersible progressing cavity pump (ES-PCP) was first used by Russians interested in efficiently running large heavy oil reservoirs containing gas for downhole petroleum production. They started to operate such reservoirs with electrical submersible PCPs.

At the same time, French companies Elf-Aquitaine and Total joined the Institut Français du Petrole now IFP Energies nouvelles through the ARTEP (Research Association for the Petroleum Techniques) to lead research into two phase pumping with high GOR. The experimental work qualified the Moineau pump which meets this requirement. Considering the volumetric characteristic of this pump type, an electrical submersible apparatus with a variable rotating speed was designed. An industrial development was managed in cooperation with the Russians who wanted to improve their own pump performance. The result was a prototype pump running in a very viscous oil well with a high GOR in Southern France in 1982.

In the eighties, Progressing Cavity Pumps had limited flow rate capacities and low head rating efficiency. Consequently, pumps driven by solid rod type drive strings, simpler and cheaper, were generally in use.

From 1995, the increase in capacity and head rating of the PCP, as well as in production of highly deviated wells with solid rod and drive head problems, led to renewed interest in the electrical submersible pump.

12.2 ES-PCP

12.2.1 ES-PCP Principle

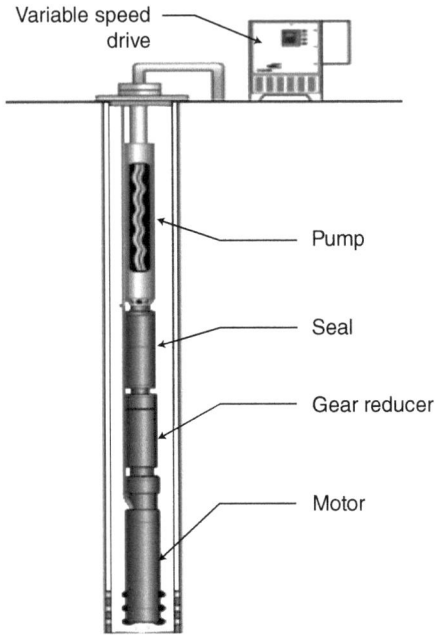

Figure 45

Electrical Submersible PCP Schematic principle.
Source: *Baker Centrilift*.

On the surface:

- Adapted well head with electrical feed through and electric power supply with variable frequency.
- The monitoring by downhole and surface sensors.

In the well from well head to downhole, the ES-PCP mainly consists of:

- Production tubing.
- Special flat three-phase electric power cable fixed by cable clamp protector and centralizer along the tubing from well head to the ES motor.
- Pump progressing cavity assembly, stator and rotor.
- Intake sub of the PCP.
- Flex-drive shaft sub.
- Protector or seal to isolate the motor and gear reducer from well fluid.
- Speed reducer gearbox to reduce the speed ES motor rotation if AS motor.

- Motor drive:
  - asynchronous type two to six pole three-phase electrical submersible motor (ES),
  - or Permanent Magnet electrical motor (PMM).

Like the ESP lift pump motor assembly that is commonly used, the ES electric motor is positioned first at the bottom. In such case the heat generated by the ES motor-gear box running is transferred by the motor housing to the flow pumped in the annulus to the pump intake. The pumped fluid viscosity is slightly reduced by the heat absorbed. If the assembly were reversed with the pump down and the motor up, the motor cooling efficiency would be greatly reduced.

12.2.2 ES-PCP Components

From surface to downhole ES-PCP components are:

12.2.2.1 Downhole Driven Rodless PCP

The pump element is run upside down and driven from the bottom compared with standard PCP driven from the surface by a solid rod. So it is important to know the rotor pump rotation direction. With the ES motor under the pump, the rotor has to rotate in clockwise direction when viewed from bottom to top or counter clockwise when viewed from top to bottom.

The PCP pump size and characteristics are adapted to the desired flow rate and pressure head required and controlled by the well production performance and the desired submergence level into the annular. The ES electric motor is selected by size, power capacities from 15 to 150 hp and the speed rotation adjusted from the surface by the variable speed drive (VSD) power supply.

Typical ESPCP are as:

- OD Size from 4" to 6.75"
- 5 to 35 stages stator
- Capable of handling up to 6.6 bar (100 psi) per stage
- ES electric motors are from 15 to 150 hp
- With 60Hz, ES motor speed is 3500 rpm (two poles)
- With 50Hz, ES motor speed is 1500 rpm (four poles)
- Gear reducer from 3/1, 9/1, 15.5/1
- Normal PCP speed range with Variable Speed Drive from 100 to 500 rpm
- Maximum intake flow 1250 m^3/d.

The PCP sub intake. The fluid flow pass from the annulus to the PCP intake has an area adapted to various fluid flow rates and viscosities.

The flex drive accommodates the eccentric rotation of the PCP rotor and the concentric rotation of the gear box shaft.

The protector or seal sub is positioned between the flex drive and the gear reducer-ES motor. It isolates the motor and gear reducer from well fluid, securing its internal cavity against any penetration of formation fluids and compensating variation of oil volume resulting from temperature variations. The protector system is a bag and seal (one or two possibly in tandem) filled with specific oil to allow the motor and gear reducer oil to expand and contract as downhole temperature varies while the motor is running. The protector also equalizes the pressure inside the electric motor with the well pressure at the level where it is hung and carries the thrust bearing load of the pump. If the protector is connected to the lower part of the electric motor, the oil compensator should complete oil leakage through mechanical fittings and those due to temperature variations.

12.2.2.2 The Gear Reducer with AS Induction Electrical Motors

Standard asynchronous (AS) induction electrical motors rotate at approximately 1,400 rpm at 50-Hz with four poles (two pairs) and 3,500 rpm at 60-Hz with two poles (one pair), but a PCP's operating range is typically 100 to 500 rpm. The relative high speed of the electric motor compared to the speed rotation of the PCP rotor requires a speed reduction between electric motor and PCP. The speed reduction, achieved using a double planetary gear designed unit, can reduce the speed by 3 to 15 times its original ES motor rpm.

For example, with 60 Hz two-pole, 3500 rpm ES motor:
– a 15.5/1 gear ratio the PCP rotor speed = 226 rpm,
– a 9/1 gear ratio the PCP rotor speed = 389 rpm,
– a 3/1 gear ratio the PCP rotor speed = 1167 rpm.

A planetary gear reducer is a complex mechanical part of the system which is more subject to failures than other parts of the electric motor and the PCP components.

12.2.2.3 The ES Electric Submersible Asynchronous Induction Motor

ES motors driving the pump are two poles (one pair) to six poles three-phase asynchronous squirrel induction type, ranging from 15 hp to 150 hp.

ES motors are filled with special oil which balances the internal pressure with annular pressure connected through the protector seal sub. The oil provides dielectric strength, lubricates the bearings, and ensures thermal conductivity. The motor thrust bearing carries the load of rotor.

The ES motor size is determined by the power required to operate the PCP depending on the fluid, the flow rate, the head of pressure, downhole temperature and the casing size.

A special three-pin connector with feed through, the "pot head", connects the ES motor to the three-phase power cable.

Corrosion-resistant coating on the pump and motor assembly can be provided for H_2S-CO_2 protection and similar environments.

12.2.2.4 The ES Power Cable

The power cable used downhole for submersible pumps is critical for overall system efficiency and reliability. Power cables are available in a wide variety of configurations to most effectively match electrical characteristics, fluid properties and wellbore geometries.

The power cable connected by a specific "pot head" connector on top of the ES motor is run up the side of the pump and strapped to the outside of every joint of tubing from the ES motor to the well head, passing through specific connectors and feed through of the well head and to surface junction box.

The cable is made in both round and flat designs depending on the well diameter for maximum clearance and minimal power lost. Cables have three solid copper isolated conductors and a metal shield to protect them from damage. Depending on downhole conditions – harsh, hot, gassy and corrosive – different types of cable are available from manufacturers. Insulated electric materials used include Kapton, EPDM rubber or polypropylene and the metal shield cable armour can be steel armour coated or galvanized. Special cables' H_2S resistance is ensured using lead sheath and K Monel armour to enhance run life.

During the set-up, special care should be given to the assembly motor/cable connector and power cable. The power cable is strapped onto the tubing with a special band it and in some cases using a specific clamp centralizer with casing fixed on the tubing string.

In specific cases with the power cable, a hydraulic control line is added for treatment, for hydraulic downhole safety valve and electric line or fibre optic line for downhole sensors well monitoring.

ES power cable has to be selected properly in order to reduce power losses across the cable and enhance overall system artificial lift performance.

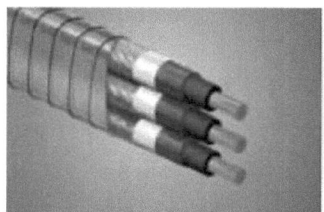  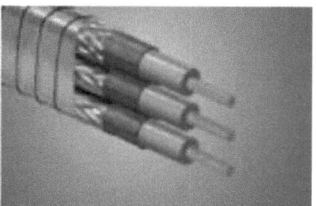

Figure 46-1

RedaMAX ES flat or round power cable.
Source: *Schlumberger*.

12.2.2.5 Cable Clamp Protectors

ES-PCP is run in the well simultaneously with the production tubing and the electric power cable located in the annulus between casing and tubing.

The most frequent cause of failure during ES-PCP start-up are damaged power cable and control lines due to crushing between the production tubing couplings and the casing internal diameter during tubing installation. Power cables and lines are expensive pieces of equipment and take time to be installed safely.

The use of **cable clamp protectors** is recommended along the tubing string to decentralize to fix and protect the flat or round ES power cable, control line and well monitoring electric or fibre optic data cable. The number of cable clamp protectors used increases with the well deviation.

Examples of cable clamp protectors:

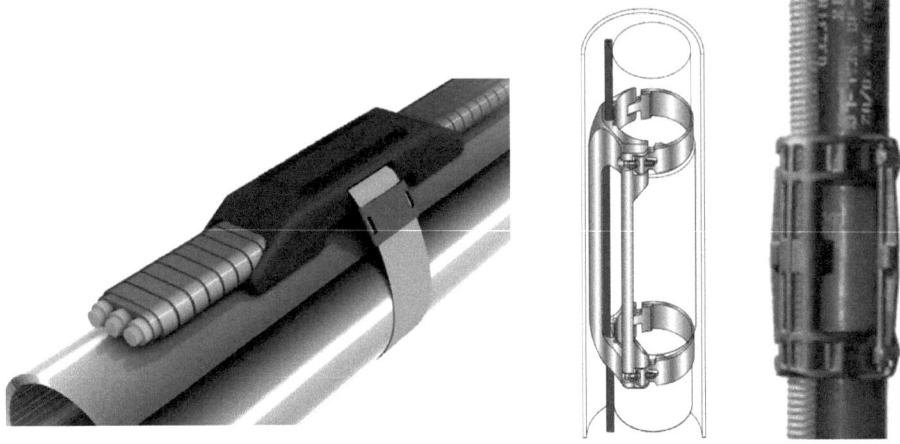

Figure 46-2

Examples of cable clamp protectors.
Composite material cable protector.   Cast steel cross coupling cable.
Source: *BossCLAMP Artificial Lift company.*   Source: *S.U.N. Engineering.*

12.2.2.6 Well Head Feed Thru and Connection

A specific wellhead with electric feed thru and connectors achieves the crossing of the power cable of the submersible motor.

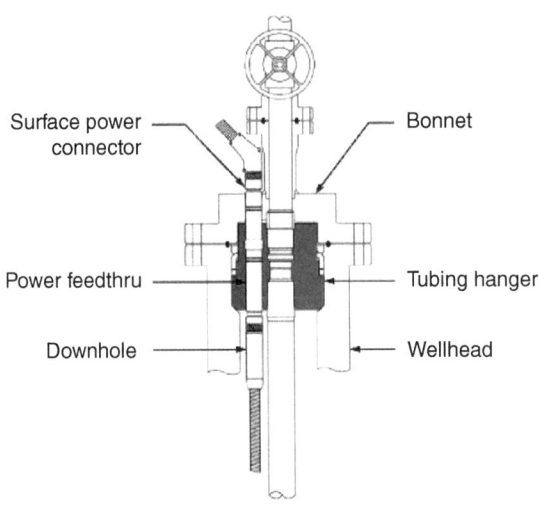

Figure 47

Well head with power feed thru connector.
Source: *ITTCannon/BIW.*

The flat or round pump cable is run through the wellhead, then connected to the well head with specific feed thru to provide a fluid block, permitting the three-phase electrical power cable to pass safely and reliably through the well's pressure barrier.

12.2.2.7 Downhole Monitoring Tool

Under the ES motor, a specific sub sensor can be fixed for well monitoring, such as pressure and temperature, and ES motor control, such as voltage, temperature and vibrations. All data is transferred to the surface *via* the power cable.

See downhole and surface monitoring chapter 15 and Variable Speed Drive (VSD) PCP's rotor, chapter 9-2.

12.3 PERMANENT MAGNET MOTORS: NEW

12.3.1 ES Permanent Magnet Motor Technology

Permanent Magnet Motors have recently appeared on the downhole artificial lift market to directly drive both ESPs and PCPs instead of standard ES asynchronous induction electric motor and gear. See chapter 9-2.7 Direct Drive Head with Permanent Magnet motor.

The ES Permanent Magnet Motor is a brushless three-phase winded stator and rotor pole with multi-permanent magnets and filled with oil. It is powered from the surface by a Variable Speed Drive (VSD) controlled by a microprocessor to continually switch the phase to the windings stator so as to keep the stator current in phase with the permanent magnets of the rotor turning.

12.3.2 PMM Advantages and Benefits

The ES Permanent Magnets Motors have many advantages and benefits compared to similar asynchronous AC induction motors.
- Higher power density, higher efficiency, less heating. The PMM requires less current for the same effect and therefore results in greater efficiency with lower heat production.
- Energy saving ranges from 10% to 30% compared to ES induction motors.
- High torque at low speed, torque decreases as speed increases.
- Smaller size OD motor envelope. The rotor dimensions decrease as permanent magnets technology develops.
- No more gear box. The PMM eliminates the complex satellite gear box, the weakest and costly components of standard ES-PCPs.
- The VSD powers the downhole PMM *via* the ES cable with lower voltage than 460 VAC compared with an AS motor.
- The VSD provides for a wide range of speed adjustment, 250 rpm at nominal torque to 1000 rpm operating range, torque control capabilities with smooth start and speed motor shut down.

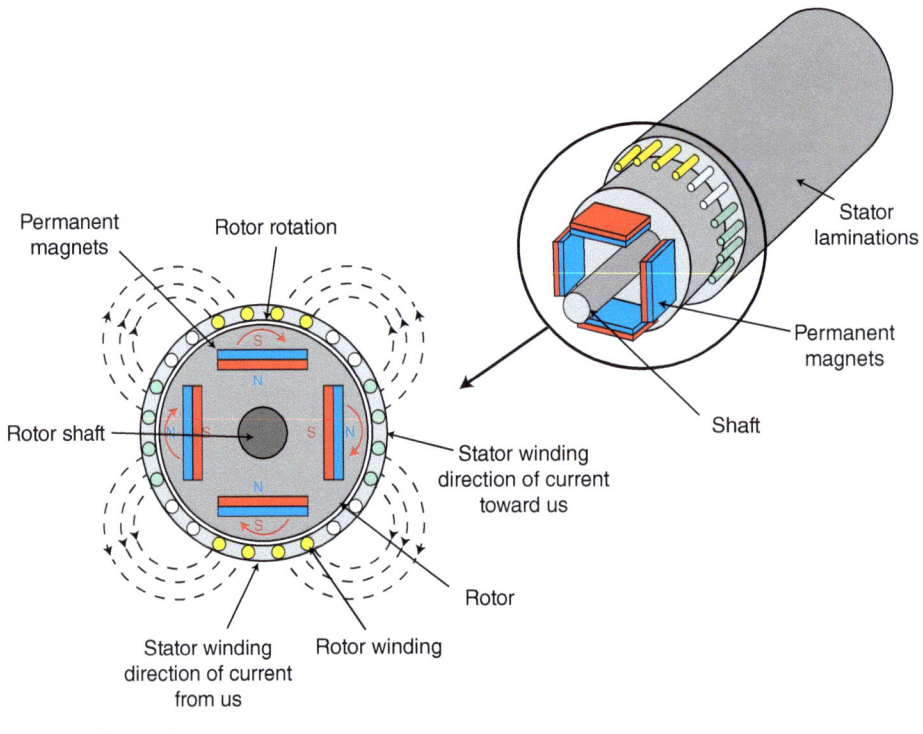

Figure 48

Permanent Magnet Electric Motor Principle.

- PM motors can operate the PCP's rotor in reverse rotation at low speeds by changing the phase sequence from two of the three lines on the surface. The pump reverse rotation back-flushes the pump. Reduced well intervention. Reverse rotor rotation is impossible from the solid drive rods.
- The VSD integrates control algorithms to increase production, improve energy efficiency and enhance the reliability of artificial lift systems.
- The PMM-PCP pumping systems can safely operate unstable wells with the speed drive control and easy start and shut down pump motor.
- The PMM-PCP is energy saving, has extended run life, increased global production and results in operating and capital cost savings.

12.4 ES-PCP RETRIVABLE THROUGH TUBING

NEW concepts to deploy, retrieve and replace downhole ESP or ES-PCPs pumps.

A retrievable ES-PCPs through tubing offers a cost saving system in high attractive cost areas.

The ES-PCP's components, such as the cable, ES motor-protector-gear and the PCP have different run-lives before failures, but one failed component generates a total pump system failure.

The purpose of the through tubing conveyed ES-PCP system is:
– To change the ES-PCP or the PCP component without retrieving the total production tubing assembly including ES power cable and downhole permanents sensors.
– To save fragile components, such as ES cable and permanent sensors if retrieved with the tubing.
– To use pump components with the expected run life.
– To retrieve pump components by using standard light lift intervention such as wire line, rods and coiled tubing units instead of the work-over rig.

The two options for retrieving the ES pump system through the tubing are:

1. Retrieve the PCP component and leave the ES motor downhole.

2. Retrieve the ES motor and the PCP components.

12.4.1 PCP's Component Retrievable Through Tubing

This option is under the perception that the weakest component of the system is the PCP's stator-rotor pump. The aim is to replace the PCP's pump component without pulling and replacing the downhole ES drive motor.

The PCP pump stator-rotor components located above the ES motor are accessible and retrievable through the tubing bore by using light intervention lift, such as wire line, rod or coiled tubing. The ES motor is secured to the lower end of the tubing string and permanently connected to the ES power cable. It is strapped to the motor and alongside the tubing to the surface well head, together with the other permanent components and sensors such as a traditional ES-PCP. The ES rotor has a specific end drive shaft extending upward to mate the driven end of the PCP rotor and a mechanical latch/unlatch system to secure both ES and PCP stator housing. On top of the PCP stator, a specific head over shot is used to fish the PCP assembly to be retrieved or installed from the surface with light lift equipment, such as a wire line, rod or coiled tubing units. The pump stator-rotor assembly can be replaced or installed in reverse.

To summarise the system: the bottom drive ES motor has the flexibility to retrieve the progressing cavity pump. These allow the operator to replace the PCP's pump assembly in a simple two trip wire line or other light operation.

"The PCP pump retrievable through tubing" is from Baker Hughes Centrilift.

12.4.2 ES Motor and PCP Retrievable Through Tubing

The other alternative to deploying, retrieving and replacing the complete ES-PCP through the production tubing is to mechanically and electrically connect/disconnect the downhole ES motor from/to the ES power cable which is permanently strapped to the tubing. The solution is to use a specific component, a downhole male electric wet connector connected to the ES power cable and located on a sub end of the production tubing. The ES-PCP's pump assembly ended with the complementary female wet connector is lowered down inside the production tubing by using light lift equipment such as wire line, rod or coil tubing units to make an electric connection with the downhole male connector to power the ES pump motor.

Principle: a downhole connector support with a mechanical latch/unlatch assembly is fixed at the end of the first tubing. The connector support contains a specific wet-mate electric connector with three separated male pin contacts connected with the three-line ES power cable. The production tubing is run down the well with the downhole permanent male connector support and with the ES power cable strapped alongside the tubing with the other permanent components and sensors such as a traditional ESPCP.

A docking sub located end of the ES motor contains the complementary three separated female electric mate-wet connectors. The whole flow is directed past the docking sub and the motor for cooling it. The pump assembly including from the bottom, the docking sub female connector-ES, motor-protector-gear and PCP ended by a fishing neck is lowered down inside the production tubing by wire line or rod or coiled tubing units. The pump assembly is latched on depth to the downhole electric connector to make the three electric continuity between both connectors. The three-phase electric lines can power the ES pump. The pump assembly can be retrieved in reverse.

The downhole wet-mate male-female electric connector is the key of the system during the all production run-life pump system because:
 - The male electric wet connector is not retrievable, it is a permanent component part of the tubing string. It has to be particularly reliable, more reliable than all other downhole components in order to be attractive.
 - The downhole male-female wet electric connector needs to have perfect insulation and continuity to power the ES motor with high voltage and high current under downhole difficult conditions in terms of temperature, fluids and gas content.

Zeitecs develops the Shuttle™ through Tubing Conveyed ES-PCP's.

12.4.3 Advantages of Through Tubing Conveyed PCPs or ES-PCPs

 - The tubing is not retrieved to retrieve and install a new ES pump system.
 - The ES power cable, source of failures and all other permanent components and sensors which are part of the tubing are run once with the production tubing.
 - The use of a light intervention unit compared with a work over rig.
 - The possible uses of a downhole well tractor and wire line to convey the pump system in a highly deviated and horizontal well.
 - Relative low cost intervention and less time required to replace the pump system compared to standard ES-PCP intervention.
 - Less lost production and faster well production recovery.
 - To adapt the pump system with the production profile and the changed reservoir condition *versus* time.
 - To carry out preventive maintenance with the use of the right pump in the right well at the right time to improve the production profile.

12.5 ES-PCP ADVANTAGES, BENEFITS AND CONSTRAINTS

Advantages:

Main advantages of ES-PCP are down the well:

In vertical deviated or horizontal installations, the main advantage compared with standard PCP rotor driven with solid rods is the fact that it is rodless. The elimination of the rod string results in operational benefits of no rod/tubing wear and eliminates problems associated with stuffing box leaks.

No rod in the tubing avoids a larger flow area in the production tubing string, lowering flow losses, increasing system efficiency and reducing the risk of the drive string or tubing failing.

Benefits:

- ESP-PCP well-suited for vertical and deviated to horizontal wells.
- High production rate possible with adapted PCP and ES motor.
- Less tubing friction loss.
- Longer life for tubing in deviated or horizontal wells.
- Eliminates tubing/rod failures.
- Not affected by sand contents mixed with the flow.
- Speed adjustment allows maximum flow rate at the lowest pressure drop.
- Elimination of the surface well head stuffing box. No seal necessary at well head compared to PCP driven with polished rod and solid rods on rotation. No leak, no seal maintenance.
- Eliminates the incorrect positioning of the rotor into the stator compared with PCP with solid rods.
- No back spin safety issues.
- No back-flush capabilities with AS induction motor. Possible with PMM used on reverse rotation at controled speed.
- Improved volumetric and overall pump efficiency *versus* standard PCP driven with solid rod.
- PCP rotor can be retrievable and replaced with slickline (specific use).
- Reduces operating expenses, less electrical power needed.
- Pump-motor-power downhole components and cable are run in the well like the ESP on standard production tubing.
- Only one tubing pull per change out.
- No maintenance for well drive head.
- Can be used with any downhole PCP manufacturer.

Constraints:

- Higher cost components as ES motor-speed reducer gear box or Permanent Magnet motor, ES power cable-cable clamp, Well head feed thru, VSD power unit.
- Mechanical limitation of complex components, particularly gear box.
- Temperature limitation for thermal recovery process. The high temperature from downhole fluid and ES motor running is applied on running components: motor, electric parts, power cable, connector and on PCP stator elastomer.

12.6 ES-PCP'S APPLICATIONS

- In deviated and horizontal wells where minimal bottomhole pressures are required for optimized production.
- In cold production wells with high viscous fluid and high sand content. In very viscous applications the elimination of a solid rod string accommodates a larger flow area in the production tubing string, lowering flow losses and increasing system efficiencies.
- In coalbed methane well de-watering operations. ES-PCP are particularly suited for applications where abrasives such as formation sands, sands used in well fracture operations and coal fines are produced with the formation fluids.
- In SAGD production, limitations are the high temperature fluids applied to the ES motor and pump elastomer. High temperature ES motors have recently become available and have been tested on SAGD production systems at 250°C (482°F). The same applies to specific high temperature elastomers (see chapter 3-3 and 13-1.4) on stator and high temperature PCP metallic stators (see chapter 13-5).

12.7 ES-PCP'S MANUFACTURERS

ES-PCP – Asynchronous Motor Manufacturers Include:

- **Baker Centrilift**: Electrical Submersible Progressing Cavity Pump "ESPCP™"

Type	Casing size inch	Motor max size OD inch	Speed range rpm	Flow rate M^3/d	Lift capacity m	Motor power kW
400	5-1/2 ++	4.5	100-500	10-60	1000-1800	12-30
500	7 ++	5.62	100-500	30-120	1000-1800	22-43
600	8.625 9-5/8	6.75	100-500	50-200	900-1800	32-80
700	9-5/8	7.25	100-500			

- **Schlumberger Reda**: Progressing Cavity Submersible Pumping System "PCSPS". Schlumberger-Reda provides the ES drive AS motor and gear.

ES-PCP – Permanent Magnet Motor Manufacturers Include:

- **Shengli** / PMM –ES-PCPump (China)

Type	Motor power kW	Speed range rpm	Flow rate M^3/day	Lift capacity m	Motor max OD inch	Casing size inch
QLB5-1/2	12-30	80-360	10-60	1000-1800	4	5-1/2 +
QLB7	22-43	80-360	30-120	1000-1800	4.8	7 +
QLB9-5/8	32-80	80-360	50-200	900-1800	6.3	9-5/8

- **Apex** rodless PC pump system (PMM-PCP) (Canada)

Apex Advanced work with suppliers to offer PCP driven by permanent magnet electric submersible motors and VSD Variable Speed Drive power supplies

- **Borets-Weatherford**: Dominium Permanent Magnet Motor (Russia)

BORETS Services – a Russian company provides PCPs with PM motors, downhole sensors and specific Variable Speed Drive power supplies.

Dominium Permanent Magnet Motors:

Type 2VEDBT motor	Motor power kW	Speed nominal/max rpm	Rated Voltage V	Current A	Efficiency %
10-117V5	10	250-1500	320	27	85
14-117V5	14	250-1500	430	27	85
21-117V5	21	250-1500	650	27	86
28-117V5	28	250-1500	860	27	86
35-117V5	35	250-1500	1080	27	86.5
42-117V5	42	250-1500	1290	27	87
49-117V5	49	250-1500	1510	27	87.5
56-117V5	56	250-1500	1720	27	88
70-117V5	70	250-1500	2150	27	88

Example: BORETS PCP with PMM motors rated at 500 rpm

Type EOVNB5	Flow rate M^3/day	Lift capacity (m)	Length (m)
10-2000	10	2000	5.85
20-2000	20	2000	5.9
30-1200	30	1200	5.6

- **Novomet** Permanent Magnet Motor for PCPs and ESPs (Russia)

Example:

- serial 319 (OD:3.19") power range 12-63 kW
- serial 456 (OD:4.56") power range 22-400 kW

Hydraulic Pump Parts stator-rotor manufacturers include:

- **PCM, Kudu**
- **Netzsch Nemo**: Electrical Submerged "NSPCP" Downhole Pump or PCP-downhole pump "ESPCP"

And others

Feed True Well Head

- BIW Connector Systems provide electrical feed thru systems

ES Cable, all ESP manufacturers
- Reda Schlumberger
- Baker-Centrilift

Cable Clamp Protector
- S.U.N. Engineering
- Winterhawk'
- BossCLAMP™, Artificial Lift Company
- Hydconline.

CHAPTER 13

New Features on PCP Pump Components

13.1 PCP HIGH CAPACITIES

The pump displacement V is a function of three parameters:
- E : Eccentricity rotor/stator.
- D : Rotor diameter.
- P_S : Stator pitch or cavity length distance of displacement of the cavity volume for one rotor rotation.

The mono-lobe (1-2) pump displacement V is the fluid volume produced in one revolution of the rotor rotation. See chapter 3-1.2.

$$V = 4E \times D \times P_S$$

The calculated pump flow rate per minute Q_c is the volume pump displacement V times the rotor rotations per minute N (rpm).

$$Q_c = 4E \times D \times P_S \times N$$

The only E, D, P_S and N parameters can be adjusted to increase the pump flow rate and high volume PCP are limited by the casing size and the drive string.

Standard PCP are driven by solid rods and it is recommended to use Hollow rods (see chapter 6.4) to drive high capacities PCP because of high-torsion rates.

High capacity PCP ranges are as follows:

Manufacturers	Model	Nominal displacement at 100 rpm		Pump lift	Pump OD		Stator length	Min. casing	Min. tubing
		M³/day	Bbls/day	(m)	(inch)	(mm)	(m)	(inch)	(inch)
R&M	15-P-334 SL	342	2154	1524	7	177	14.18		
Baker	1400-G-1300 223-G-90	223	1400	900	5.125	130	11.8		
Kudu	200-K-1450	200	1260	1450	5.43	138	14.15		
Nov Monoflow	200-1800	200	1260	1800	7	177	10.2		6 5/8
Oillift	200-1200	200	1260	1200	5	127	15.68		
PCM	20E860 100TP860	160	1000	860	5	127		6 5/8	4
	185-E-1500 900-TP-1500	185	1156	1500				8 5/8	4 1/2
Protex	18-PK-175	175	1100	1800	5.125	130			
Europump	120-E-11.5 175-E-7.5	120	750	1150	3.937	100	14.7		
	175-E-7.5	175	1010	750	4.5	114			
Netzsch	NTZ 550 100ST 145	145	912	1000	5.5	140			
Moyno	9-H-150	150	950	900	4.5	114	14.9		
Dyna-Lift	DD-98-1600	98	616	1600	5.5	140			

13.2 PCP MULTIPHASES FLUIDS

By design, the PCP volumetric pump is capable of handling solid, liquid and vapour phases. Standard PCP design can handle up to 40% of gas at pump intake, but are not designed to pump multiphase oil and water mixtures with high gas void fraction (GVF) content.

With standard PCP in the well, when the gas void fraction of the multiphase fluid exceeds 40%, the gas compression and pressure distribution inside the PCP pump leads to temperature elevation. The reduced liquid fraction generates insufficient cooling. The result is a rapid increase in rotor-stator friction, an excessive build-up of torque and heat, a loss of efficiency and a reduced life span. In the worst case, the result is premature damage to the elastomer pump's stator or its failure.

Surface monitoring parameters and alarms can detect excessive drive torque and power consumption with lower to no surface flow rate. In such cases, the PCP surface drive head has to be stopped.

PCP Multiphase with Hydraulic Regulated from PCM (HRPCP)

The multiphase progressing cavity pump, HRPCP (Hydraulically Regulated Progressing Cavity Pump) from PCM Moineau is capable of handling multiphase fluids of up to 90% gas. The HRPCP concept is patented, and was invented by C. Bratu.

The HRPCP pump is a Progressing Cavity Pump with a standard stator and a modified rotor. The rotor includes a system of Hydraulic Regulators (HR) installed between the cavities along the pump length. The HR concept is a series of circular channel drilled hole ports inside the rotor. Each hole port is sized to transfer fluid volume and pressure between two successive cavities. In order to better control the HR regulation process along the rotor length, valves or nozzles can be added in each port.

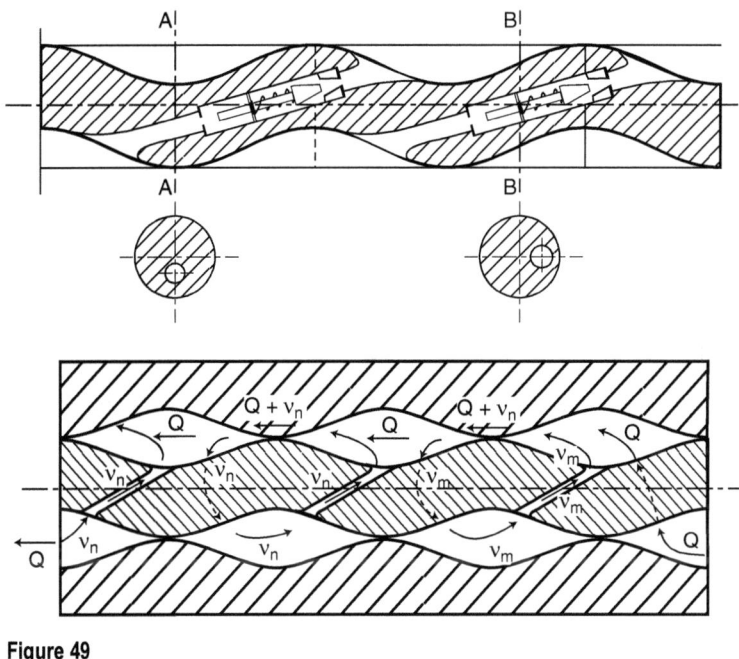

Figure 49

Multiphase PCP rotor and stator principle.
Source: C. Bratu patent FR 2865781, US 7,413,416B2.

Along the length of the pump's hydraulic regulators is a fluid regulation process.
It make it possible to:
- Re-circulate fluids between two successive stator cavities in order to compensate for the compressed gas volume between these cavities.
- Control the balancing pressure between cavities and uniform the pressure distribution and the temperature across the pump length.
- Homogeneously distribute temperature so as to avoid an excessive build-up of heat.

Compared to a standard PCP, the multiphase HRPCP pump results are a lower stator temperature, less stator strain, lower friction torque and less power consumption. It protects the stator and therefore improves the pump's performance, resulting in a longer run life.

PCM provides HRPCP pumps with a specific hydraulic regulator design based on determined well data.

13.3 PCP AND THERMAL HEAVY OIL RECOVERY

The thermal process is a way to produce heavy oil.

Common and proven heavy oil thermal recovery processes are Steam Assisted Gravity Drainage (SAGD), Cyclic Steam Injection (CSS) and Continuous Steam Injection (CSI) or Steam Flood. Thermal methods involve the injection of steam or hot water into the reservoir to assist the heavy oil recovery, reduce the oil viscosity, improve its mobility and provide fluid displacement mechanisms to the well producer.

Three thermal production methods are:

- Steam Assisted Gravity Drainage (SAGD): Using pairs of parallel horizontal wells: the top well is the steam injector and the other, the bottom well, is a producer. The steam chamber from the top well heats the oil, reducing its viscosity so that it can drain down by gravity into the lower producer well.
- Cyclic Steam Stimulation (CSS) or "huff and puff" technique. A single well is used to inject steam into the reservoir, both to heat the oil and to reduce its viscosity. After the reservoir has been through a steam injection phase, the operation is reversed to produce the heated oil.
- Continuous Steam Injection (CSI) or Steam Flood: Two vertical wells are used. One injects steam into the reservoir, creating a steam front that sweeps the oil before it into the producer well.

The heated heavy oil producer is not eruptive and artificial lift techniques are used to produce the heated oil. Methods used are gas lift, ESP, PCP and ESPCP.

The artificial lift technique used is an important part of the success factor to produce thermal heavy oil recovery because of the high volume lift (oil, water, steam, vapour gas) and the recovery factors, but thermal heavy oil recovery involves high capital expenditure and operating costs.

Main lift difficulties are the fluid temperatures and the multiphase content.

Main constraint with high temperature fluids is the reliability *versus* time of the downhole pumps components such as elastomer used for seals, elastomer PCP stator, electric motor, shaft bearing and the well completion.

PCP specific technologies to pump high temperature mixed oil and vapour fluids are such as:

- PCP stator made of high temperature elastomer resistance.
- PCP with metal stator.
- PCP and steam injection through rotor.

13.4 PCP HIGH TEMPERATURE ELASTOMER STATOR

Elastomer technologies are the central element of the PCP stator and the key of the PCP run-life as described in chapter 3, 3-7.

Standard elastomers currently available on the market are operating temperature limited to approximately 140°C (280°F). This is generally high enough for many wells with down-hole PCPs.

For thermal heavy oil recovery, manufacturers provide proprietary specific elastomers, for example:
- Robbins & Myers, Moyno HTD350™ Down-Hole Pump.
  Compatible with steam injection, capable of handling down-hole temperatures of up to 150°C (300°F). The elastomer is mechanically secured to the stator tube with no bonding agent.
- Kachelle (PCP Even Wall).
- Weatherford (Uniform-Thickness Pumps) High-temperature elastomer up to 572°F (300°C) allows stator to remain in well during steam injection.

13.5 PCP WITH METAL STATOR

Conventional elastomer PCP stator has a negative to zero clearance with the metal rotor for making a seal and to reduce the leakage between cavities. A new concept, the metal stator, eliminates the elastomer stator material.

13.5.1 The Metal Stator PCP Principle

A Metal-to-metal progressing cavity pump "M-PCP" is based on a single helical metal rotor rotating inside a double helical metal formed stator. A positive to close to zero clearance is fixed from the manufacturer process between the metal rotor and the metal formed stator. The clearance between the two metal elements permits rotation and basically eliminates wear between the two metal components. But clearance allows leakage or slipback due to suction, which decreases the net output flow rate for a given pressure rise. To reduce the leakage or slipback, the rotor speed needs to be increased up to 500 rpm compared to standard PCP with elastomer stators.

Past Experience, First Tests from PDVSA

A PCP with metallic casted stator has been developed in the past by PDVSA and tested in Venezuela. Stator pumps were manufactured by element casting and different pumps where tested in a loop with encouraging results.

Publications: A.Olivet, J.Gamboa, 2002, *"Understanding the performance of a progressive cavity pump with metallic stator"*. SPE 77730 (presented at the 2002 SPE ATCE conference, San Antonio, 29 Sept-2 Oct).

13.5.2 Metal Stator: M-PCP Vulcain™

PCM pump manufacturer has developed the Vulcain™ M-PCP pump with a metallic stator. The Vulcain™ M-PCP describes the design wherein the stator is made of two parts. The first part, in contact with the rotor, is made of a thin tube formed from a low-elasticity material like metal, and the second is designed to apply and maintain stress exerted by the first part on the rotor providing controlled zero to small clearance for the required pumped fluid pressure gain. Current difficulties involve manufacturing the metallic stator sleeve with tight and constant tolerance.

A prototype PCP with a metallic stator has been tested by PCM, developed and marketed under the name Vulcain™ M-PCP.

PCP Vulcain™ pumps are well suited for high-temperature fluids to produce heavy oil with injection SAGD and cyclic steam stimulation processes.

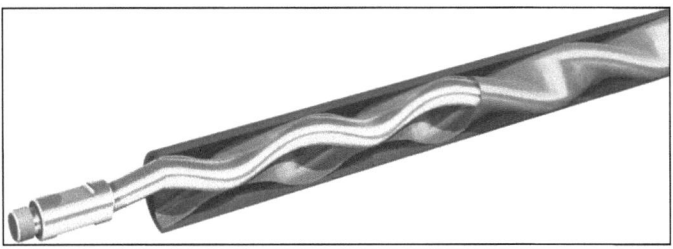

Figure 50-1

PCP pump metal stator and rotor.
Source: *PCM Vulcain™*.

PCM Vulcain™ publications: Bauquin J-L and all dated 2005 SPE 97796, JPT May 2006, MEALF 2009 Barhain. WHOC 2008-498 Canada, WHOC 2011 Canada.

From PCM, the characteristics of the Vulcain™ pump are:
- Metallic stator, metallic rotor
- Max. operating temperature : 400°C (750°F)
- Max. head: 100 bar – 1450 psi
- Very efficient at low submergence.

The 550 MET 750 can produce 550 m^3/day at 0 bar and 500 rpm.

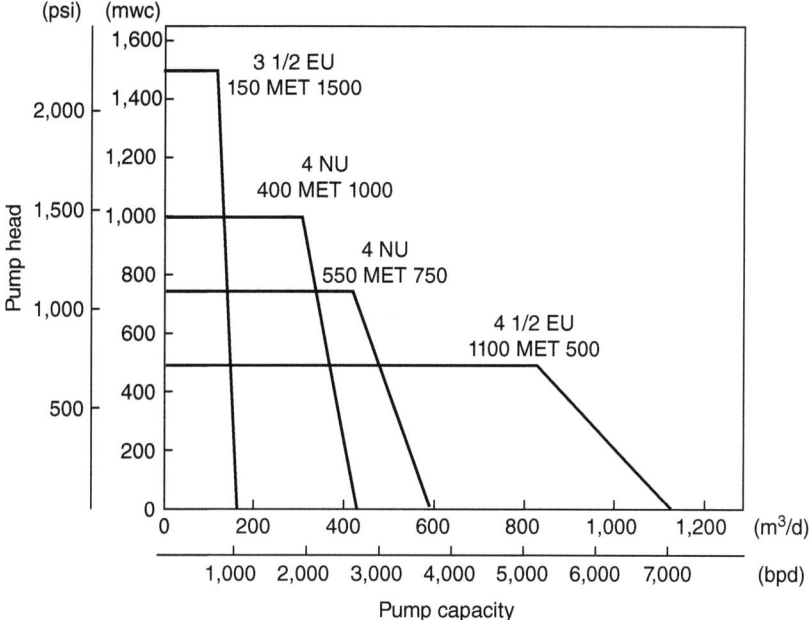

Figure 50-2

PCM Vulcain™ MET Pump characteristics.
Source: *PCM*.

13.5.3 Others Metal Stator PCP Manufacturers

- **C-Fer Technologies** carries out research into Metal-PCP design and has been successfully operating in Cuba since 2005. Published by PC-PUMP New Development metal-to-metal PCP pumps (JPT July 2009, SPE 120645).
- **Robbins&Myers** Energy Systems has developed a downhole one-piece metal to metal rotor/stator PCP's pump designed for 1200 m of lift.

The Moyno HTD660™ one-piece metal to metal rotor/stator can operate in temperatures up to 350°C (660°F). An additional advantage is its ability to circulate steam through the stator in Cyclic Steam Stimulation (see chapter 14-1 PCP for heavy oil recovery steam injection process).

13.5.4 PCP Metal Stator Patents

- **IFP patent**, H Cholet, E Vandenbroucke, "Progressive cavity pump with composite stator and manufacturing process" FR 2,794,498 filed 07 June 7, 1999, US 6,336,796 Jan. 8, 2002.
- **PCM patent**, Lionel Lemay "Method for making a Moineau stator and resulting stator" FR2,826,407 filed June 21, 2001, US 6,872,061 March 29, 2005.

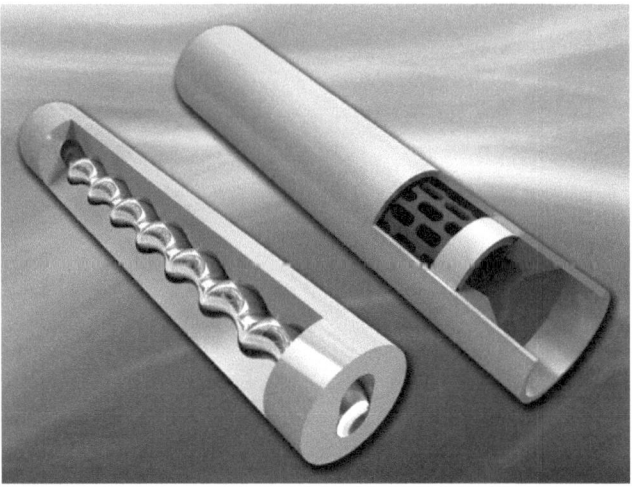

Figure 51

Solid metal stator.
Source: *Robbins&Myers Energy Systems*.

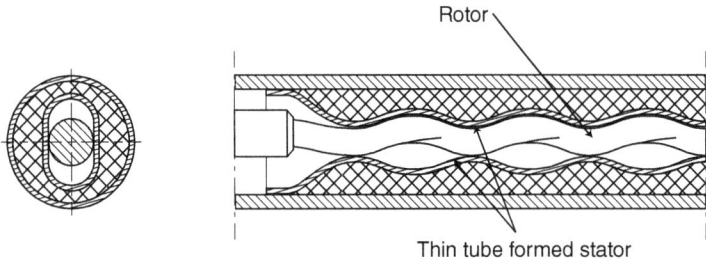

Figure 52

PCP metal stator principle patent H.Cholet E.Vandenbroucke.
Source: *IFP Energies nouvelles*.

13.5.5 PCP Metal Stator Benefits

The Metal Stator PCP technology has the benefits of standard progressing cavity pumps, but extend the application ranges of PCP for higher temperatures fluids.

The Metal Stator PCP eliminates the chemical and thermal degradation observed on current nitrile-lined elastomer stator elements, raising the maximum downhole temperature limit up to 350°C-660°F.

The Metal Stator PCP is suited for thermal enhanced oil production for handling hot oil encountered during thermal recovery methods, such as steam-assisted gravity drainage (SAGD) and cyclic steam stimulation (CSS).

13.6 PCP UNIFORM THICKNESS STATOR ELASTOMER

By forming the metallic stator tube, the shape of the stator housing is similar to the inner shape of the stator. In such cases, the thickness of the elastomer is consistent throughout the stator. These types of pump are known as Uniform Thickness Stator Elastomers.

The lower elastomer volume with the greater uniform wall thickness distribution of elastomer inside the stator tube minimize internal stator profile distortion during the elastomer injection process. This results in more uniform fluid flow between the rotor and stator surfaces.

This uniform thickness stator design results in benefits.

- Uniform elastomer thermal expansion, reduced heat build-up and improved heat dissipation.
- Elastomer distortion caused by heat and chemicals is avoided.
- Swelling is minimized and the fitting shape between rotor and stator is optimized.
- Less wear, longer service.
- High pressure/high temperature (HP/HT) tolerance.
- Increased pumping efficiency.
- Reduced pump operating torque and lower drive power for the same rate and the same height.
- Increased pump operation life.

Suited to handle the most demanding environments with proven performance for high API applications in light to heavy oil, thermal, dewatering and coal bed methane operations.

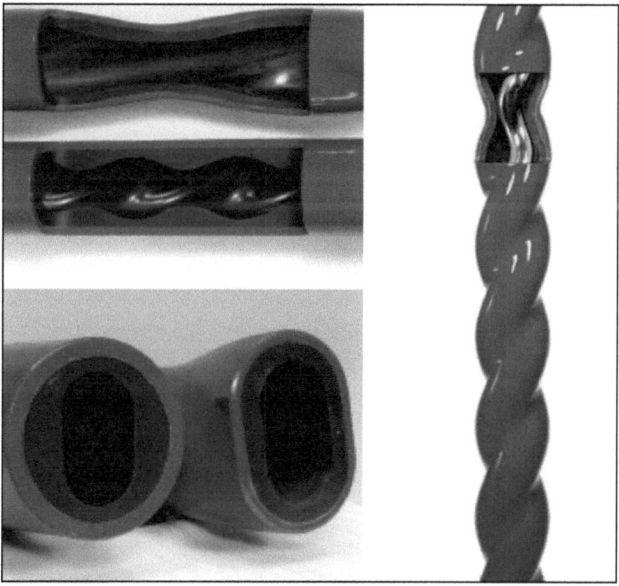

Figure 53

Uniform thickness stator elastomer and standard stator.
Source: *EUROMax PC pumps*.

PCP manufacturers & suppliers which provide uniform elastomer thickness stators are:
- Kaechele: "the Even Wall PC Pump system"
- EUROMax PC pumps,
- Weatherford: the "Uniform-Thickness Pumps, Even Wall® stator"
- CANAM
- Netzsch: the "NTU Thickness Uniform stator"
- R&M Energy Systems: the "Even-rubber-thickness stator"
- KUDU: the EvenWall™.

13.7 PCP WITH HOLLOW ROTOR

The PCP **hollow rotor** is a rotor with a uniform contour wall thickness manufactured with a special forming process.

The advantages of a hollow rotor are:
- A uniform contour wall thickness, lighter than a solid rotor.
- For single lobe and multi lobe.
- Better interference rotor-stator fit.
- Allows an inner passage within which fluids can circulate through the PCP rotor.

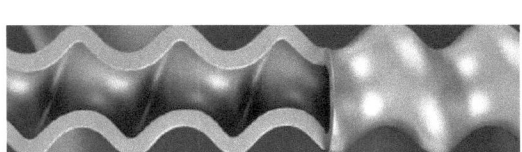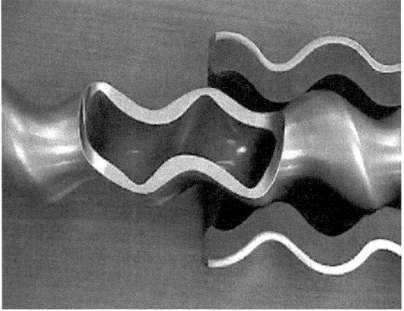

Figure 54

Hollow rotor. The *Even Wall®* Rotor.
Source: *Kachelle*.

13.7.1 PCP with Injection of Diluents in the Drain Producer

New concept: a length of small pipe is fixed onto the end of the hollow rotor to the horizontal section. Diluent injected from the surface is directly mixed with heavy oil in the horizontal drain. The mixed oil and diluents with reduced viscosity flow more easily to the pump inlet. The recovery of heavy oil is increased.

Referring IFP patent C Wittrisch: FR 2 859 753 filed 16/09/03, US 7.290.608 Nov 6, 2007.

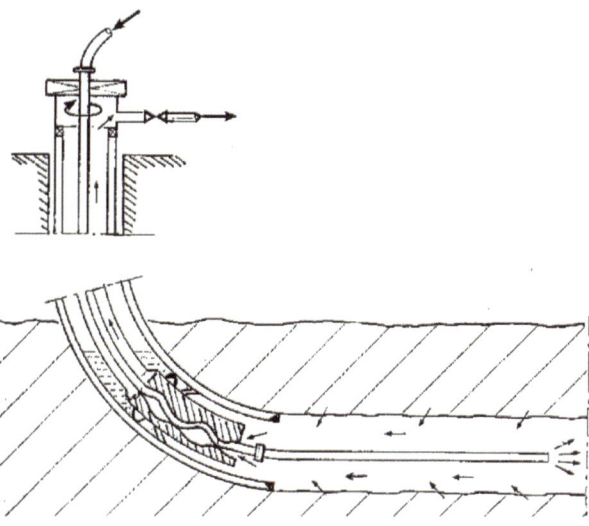

Figure 55

Principle: injection through rotor to the horizontal drain hole.
Source: *IFPEn patent figure.*

13.7.2 Adapted Drive Head for Injection Through Hollow Rods

Diluents or steam are injected down the well by using hollow rods or coiled tubing and a specific fluid injection swivel located above the stuffing box of the adapted drive head or on top of the hollow drive head.

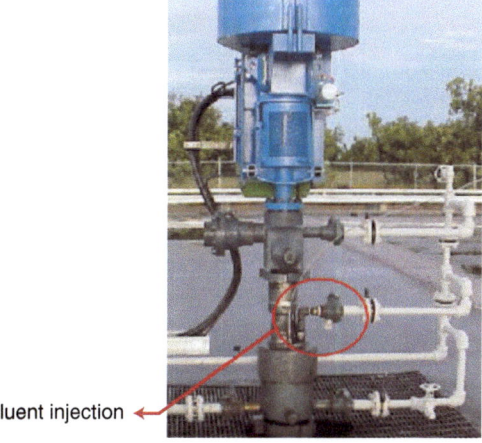

Figure 56-1

Drive head with a lateral injection swivel port.
Source: *Kudu.*

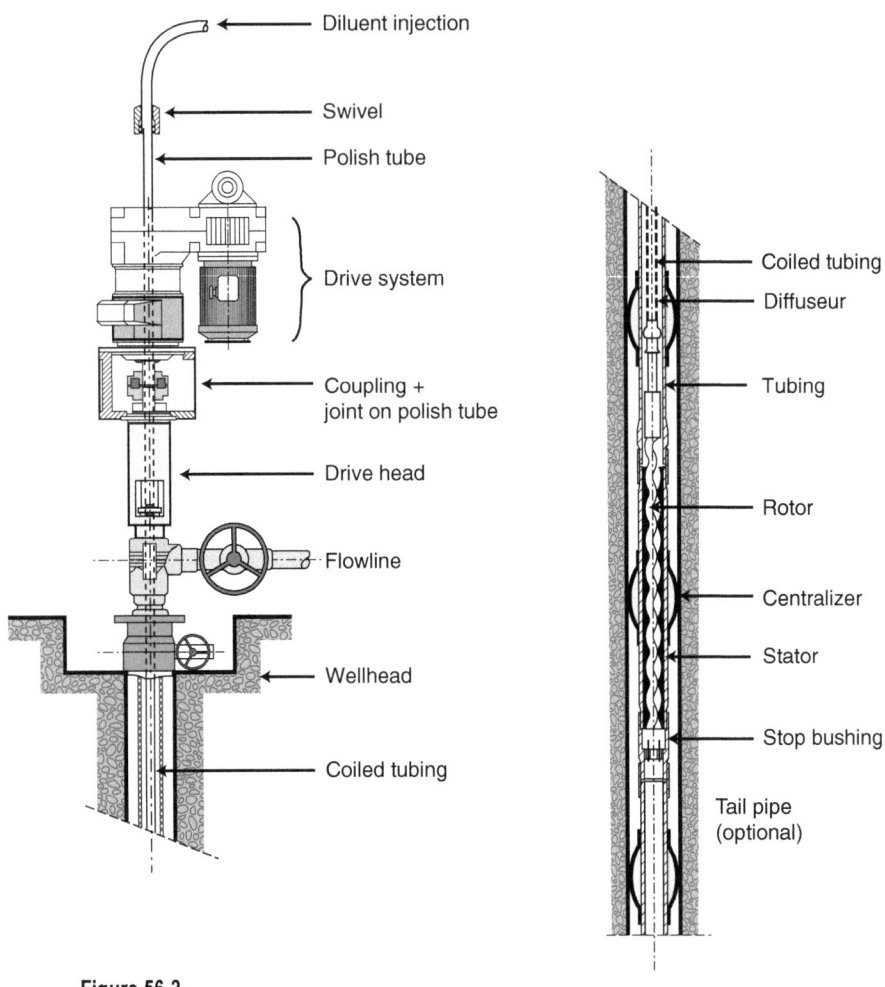

Figure 56-2
Hollow drive head injection and coiled tubing drive string.
Source: *IFP Energies nouvelles*.

13.8 PCP WITH METAL OR CERAMIC COATED ROTOR

Different types of new coating for PCP rotors are being offered by manufacturers such as:

13.8.1 Spray Metal Coating Applied on PCP Rotors

Some PCP pump manufacturers market specific steel rotors with spray metal coating instead of the standard practice which involves typical electroplating chrome deposit. The metal coating can be applied to all base alloys. It is a thin dense composite carbide material uniformly distributed over the whole the surface area. Metal coatings are now used in the industry with applications on pump components, ball/seats, barrels and plungers, valve rods and bushings, threads, gears and bearings.

Metal coatings are resistant to high temperatures (up to 775°C) and have high resistance to corrosion and abrasion.

In downhole corrosive environments in light, medium and heavy oil, metal coatings on PCP rotors provide significantly improved resistance to corrosion and abrasion.

The result is a reduced friction coefficient with reduced wear and less deterioration of efficiency over time.

The benefits include an extended rotor run life with rotors needing to be replaced less frequently than standard electroplating chrome rotors and lower operating expenses.

13.8.2 Ceramic Coating Applied on PCP Rotors

Ceramic coating is another technology commonly used in the industry on many types of metal and components. It can be applied on PCP rotors.

Scaling particles are deposited inside the PCP on the elastomer and the rotor surface can be problem. It increases the torque, load oscillation and the rod fatigue failures. Scaling particle deposits are more common on chromium coating rotors than on the inner surface of stator elastomers.

PCP rotors with ceramic-coated surface have been successfully tested and are now being marketed.

The advantages of ceramic-coated rotors include:

– Less scaling deposit on ceramic coated rotors than on electroplating chromium rotors.
– The hardness of the ceramic coating is much greater than the chromium coating.
– The ant wear rotor property with ceramic coating increases the PCP run life.

An optimum fit has to be made between the ceramic-coated rotor and the stator elastomer rigidity.

The result can be a lower torque, reduced load oscillation and rod fatigue failures, electrical power savings. The reduced wear results in less degradation of efficiency over time, an extended rotor run life and benefits for producers with lower operating expenses and higher production at high rates.

PCP specific coated rotor manufacturers are:

– Kudu: specific rotor with spray metal coating, the patented "Tough Coat™".
– Eagle Innovations Inc, St. George, Utah provides the "E"-Carbon rotor metal surface coating.

13.9 VARIABLE SPEED DRIVE AND PCP

For application to surface and downhole ES motors to drive PCP rotors.

13.9.1 Why a Variable Speed to Drive PCPs?

Problems can occurs with NO variable speed system to drive the PCPs, because standard asynchronous AC motors immediately proceed to running speed to drive the PCPs. During initial start-up at running speed, a high fluid level in the well causes the pump to produce at a greater efficiency due to increased pump suction pressure. This allows the accelerated fluids surrounding the well bore to surge, in some cases bringing unwanted solids into the PCP pump where they can become trapped. Trapped solids can often be the cause of a complete lock down of the rotor-stator pump whereby the PCPs must be removed from the well, repaired at the surface, and then relocated before operation may recommence.

Furthermore, as the oil level descends, the pressure differential between downhole and surface increases, requiring more horsepower to produce the same amount of produced fluid. Because AC motors run at constant speed, they are unable to compensate for falling oil levels to efficiently produce and maintain a desired fluid level and production rate.

13.9.2 Electric Variable Speed Drive PCPs

The Variable Speed Drive (VSD) powers and controls the surface or downhole electric motor used to drive the PCPs. The VSD is commonly a variable frequency three-phase power in the range of 30 Hz to 90 Hz maximum type. It can be used to adjust the speed of the motor and pump to fit the expected production as downhole conditions change.

In addition to VSD, intelligent automation based on real-time well dynamics controls and protects by programmed alarms. VSD also limits the pumping system by changing lift conditions, such as flow or pressure rate to optimize production and reduce risks.

VSD manufacturers. Many manufacturers provide VSD to power and control the surface or downhole PCP electric motor.

Examples: ABB, Baker-Centrilift-Electrospeed-microDRIVE, NOV Monoflow, R&M – the Guardian, Novomet, Reda-Schlumberger, SpeedStar MVD-Varistar, Borets-Axion VFD, WG Wood Group.

13.9.3 Hydraulic variable speed drive

A hydraulic surface power unit provides hydraulic fluids at a desired pressure and rate. It can be used to adjust the hydraulic motor speed located on the surface or downhole to drive the PCP.

13.10 DOWNHOLE "HYDRAULIC MOTOR" TO DRIVE PCP

The "Hydraulic Progressing Cavity Pump" H-PCP is a new concept which uses a hydraulic motor to drive the downhole PCP's rotor instead of the ES electric motor as is the case with the ES-PCP (see chapter 12).

The hydraulic power fluid generated on the surface is transferred to the downhole hydraulic motor with two coiled tubing lines. The hydraulic fluid (pressure-flow rate) power the downhole hydraulic motor to drive the PCP's rotor at variable speeds of 10 to 400rpm with controlled torque.

The hydraulic progressing cavity pump's advantages compared with the ES-PCP is the lack of electric components such as the ES motor to drive the PCP's rotor and no electric cable to power the motor.

The hydraulic progressing cavity pump can be used for all types of vertical to horizontal oil and gas well applications.

13.10.1 The Hydraulic Lines

To power the downhole hydraulic motor, two to three specific encapsulated (1"-25.4 mm) coiled tubing lines, the "FlatPak", have been developed by CJS coiled tubing Lloydminster Alberta Canada. From the basic principle one line is used to transfer the hydraulic power fluid from the surface to the downhole motor and the second line is used to return the same fluid to the surface hydraulic power generator.

A capillary injection line and data lines to the sensor wire can be added in the "FlatPak". The capillary injection line can be used to directly inject chemical friction reducers and corrosion inhibitors or diluents into the PCP's intake. The real-time downhole data line transfers monitoring data from the downhole sensor such as pressure, temperature, discharge pressure and backside pressure.

Figure 57

Hydraulic lines FLATpak™ to hydraulic drive PCPs.
Source: *CJS*.

13.10.2 Advantages of the PCP Hydraulic Drive

Compared to standard PCP:
- No surface drive head and stuffing box seal compared to PCP driven with polished rod and solid rods on rotation. No leak, no seal maintenance.
- Eliminates rod and casing wear.
- Longer life for tubing in deviated or horizontal wells.
- Less tubing friction loss.
- Larger flow area in the production tubing.
- H-PCP well adapted for vertical, deviated to horizontal well.
- Hydraulic PCP can be laid down at a 90-degree deviation in the horizontal section.
  - No more hydrostatic column left which limits the overall production.
  - The maximum oil potential of the wellbore is recovered from the entire hydrostatic column.
- Changing speed is possible, allowing maximum flow rates at lowest pressure drop.
- Reduced servicing and down time.

13.10.3 Compared to ES-PCP

No electric components downhole such as ES motor, No electric cable and feed through well head.

Reference US Patent Joe E. Crawford, application US2007/0253843 – Hydraulically driven oil recovery system.

The NEW HP-CP is still in junior but promising market.

CHAPTER 14

PCP: Many Uses

14.1 PCP AND HEAVY OIL PROCESS

Heavy oil at reservoir conditions are characterized by high densities in the range of 12° to 20°API range and high viscosities of 1,000 cP or more. The low downhole pressure does not allow for production *via* natural flow from downhole to surface. To enhance heavy oil reservoirs, water drive or gas drive are not favourable since the mobility of water or gas is too high compared with the oil mobility.

To produce heavy oil from a reservoir at an economical rate and improve recovery, several techniques have been developed, such as horizontal well architectures and artificial lift.

PCPs are the most common artificial lift to enhance thermal and cold recovery process.

14.1.1 PCP and Thermal Production

Steam Injection Recovery Process

The use of hollow rods and hollow rotors makes it possible to inject steam from the surface into the inlet PCPs.

The concept uses only one well for the injection and production. The steam provided from the surface is injected through the drive head, the tubing rod string and the specific PCP rotor. The steam injected builds up a steam chamber that rises to the top of the reservoir to heat up the oil, improving the flow characteristics, hot oil drains down into the production well.

The purpose is to reduce the downhole oil viscosity at the inlet pump and the pressure lost in the production string.

The hollow rod string or coiled tubing is used to drive the PCP hollow rotor and to inject steam or chemical diluents under the pump.

An improved PCP with high-temperature stator elastomer or a metal stator allows to remain the pump in the well during steam injection.

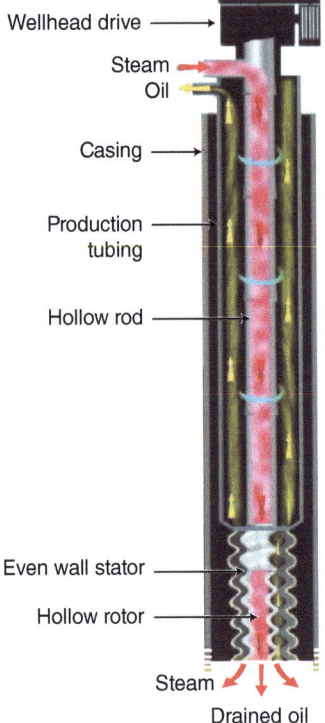

Figure 58

Steam injection through the PCP hollow rotor principle.
Source: *Kachele*.

14.1.2 PCP and Cold Production

PCP and injection of diluents recovery process.

The cold production recovery process makes use of various types of diluent injected downhole to reduce heavy-oil viscosity. Diluents such as light oil, kerosene or naphtha are injected into the drain and naturally mixed with heavy oil. The use of diluents enables heavy-oil viscosity to be reduced from 100,000 cP to 1,000 cP or less (see chapter 4.3). The diluted mixture is then lifted to the surface by the PCP.

The injected diluent dilutes the heavy oil and reduces the pumped fluid viscosity. The aims are to increase the oil recovery and to reduce the tubing pressure lost, the pump power and energy saving to lift the mixed fluid to surface.

Different Options are Available for Injecting Diluents

Option 1 – Injection of diluents independently of the PCP

The Petrocedeno (ex Sincor) project is an example of production *via* dilution of heavy oil where the injection line is a coiled tubing or a tubing string type 1.6" to 2" located in the annulus between the casing and the production tubing from the surface to the pump level.

The injection line is extended underneath the PCP pump *via* 2-3/8" tubing located inside the slotted liner (or sand screen) up to a point close to its end.

The production string is lowered down in a single operation including downhole to surface tubing injection, high volume PCP, production tubing and injection line. The diluent used is 47°API naphtha injected at the bottom of the slotted liner. It moves slowly in the horizontal section of the drain *via* the effect of the pressure differential generated by the pump. The heavy oil progressively mixed with naphtha arrives at the PCP inlet with a reduced viscosity that is acceptable for the pump. For example, under downhole conditions at 50°C of 10,000 cP, 7.5 to 9°API heavy oil mixed with 20% of 47°API naphtha, has its viscosity reduced to 200 cP. The viscosity of the mixed pump fluid can be easily adjusted from the

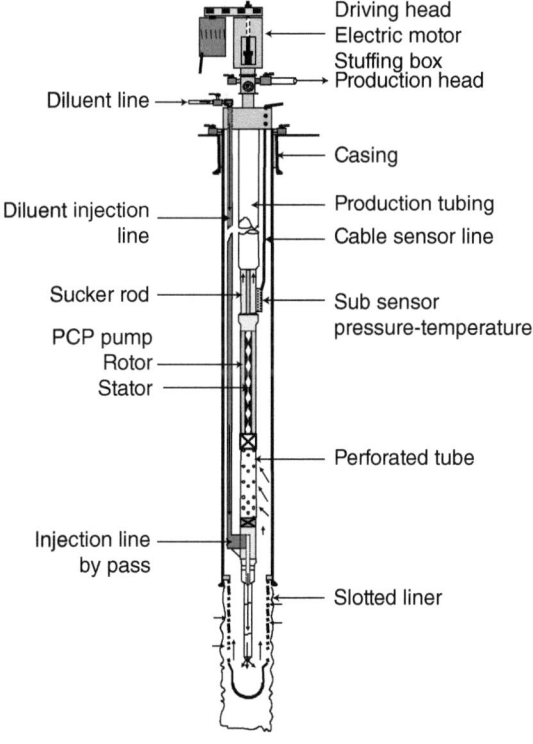

Figure 59

Cold production with diluents through injection line.
Source: *IFP Energies nouvelles*.

surface *via* the rate of the diluents injected downhole through the injection line. The production rate can be up to 1,000 m^3/day (6,300 bop) of diluted oil. The heavy oil produced by the diluents is upgraded to 32°API Syncrude oil.

Option 2 – Injection of diluents through hollow rod drive and rotor

The PCP rotor driven from the surface by an adapted standard tubing or hollow rods or coiled tubing makes it possible to inject fluids (e.g. diluents) from the surface at the PCP's level.

- With a PCP solid rotor the injected diluents port can be positioned above the PCP to dilute and reduce the pumped fluid viscosity to reduce the loss of pressure, pump power and energy to lift the pumped fluid and diluents to surface.
- With a specific PCP hollow rotor, (see chapter 8-4, 13-7) the diluents flowing near the pump suction are naturally mixed with heavy oil, making it possible to reduce the fluid viscosity at the pump inlet and pump power.

Kudu market coiled tubing drive PCP's (see chapter 13.7.2).

14.2 PCP AND WATER FLOODING

Mature water flood producing wells necessitates a larger lift system to produce high volume fluids. While the percent of oil in the produced fluid decreases, the total volume lifted must be increased. High Volume PCPs (see chapter 13.1) are an economic alternative to high lift WOR wells with less electrical usage when compared to beam and ESP systems.

14.3 PCP FOR DEWATERING GAS WELLS

Many natural gas wells, particularly at the end of their useful life and from shale and coalbed bed methane reservoirs, do not have enough flow and/or pressure to lift the water that collects at the entrance of the well.

Dewatering natural gas wells allows recovery of the low pressure gas remaining in the reservoir, however, due to marginal economic considerations, the process of removing the water must be carried out at very low cost. The PCP pumps are particularly suited to lifting the water produced by methane gas desorption to enhance gas production by lowering the downhole hydrostatic pressure productive zone. In many cases, the formation water produced is charged with abrasive particles, such as sands used in well fracturing operations and from coal fires. For water/methane applications, PCP manufacturers provide special resistant elastomer stators with reduced abrasive wear and swell normally induced by water.

CHAPTER 15

Monitoring & Controller Well Production

15.1 OVERVIEW

Modern monitoring systems is the ability to remotely measure pump performance and change pump activity using two-way communication control devices.

These "intelligent" systems can measure the performance of a single well or an entire field using sensors and data transmitting devices that phone field personnel when pump performance has changed. By comparing measurements of pump performance over time, the operator can analyse and optimize the operation of the PCP and the well itself, thus improving efficiency, and better managing field operations. Modern monitoring and controller systems are increasingly applied to wells that use PCPs.

15.2 LIMITATIONS, CONSTRAINTS, BENEFITS

Progressing cavity pumps provide a number of features and multiple constraints on equipment and reservoirs which make it possible to control production easily.

PCP pump system limitations come from surface equipment, the drive mechanism, the motor, speed, torque, thermal capacity, the electric power supply, voltage-current-frequency and on downhole, the rod string characteristics, length, torque and the pump element, size, stator elastomer characteristics, flow rate, pressure inlet-outlet and dynamic fluid level while the pump is running.

Reservoir specification capacities and production constraints include the fluids' characteristics and chemical components, the inflow conditions, often variables such as coal-bed methane, high gas/oil ratio, thermally stimulated wells, dewatering shale gas and the well completion.

15.2.1 Why Monitor Wells

PCP artificial lift monitoring, controller and real-time data with advanced diagnostics tools and analysis help the operator:

– to remotely identify potential problems,

- to take preventive or corrective action to optimize pumping systems,
- to adjust the pump's characteristics to match the well's operating conditions,
- to improve energy efficiency and savings,
- to prevent pump errors or unplanned shutdowns.

Continually monitoring and comparing parameters such as electricity usage, surface drive unit speed, rod string torque, fluid movement, pump intake and discharge pressure, and the bottom hole pressure help the operator to detect problems occurring such as paraffin build-up, perforation scaling, sanding, pump starvation, formation skin damage, and surface flow restrictions.

15.2.2 Benefits

Intelligent downhole PCP monitoring and surface real-time data control and modelling make it possible to:
- closely monitor performance between wells and pumps,
- maximize the overall production by adjusting all these interdependent parameters and limitations to operate at the constraint limiting production at each instant of time without compromising efficiency or reliability,
- automatically notify the operator when key parameters vary beyond specified limits,
- maximize production and add an overall life cycle pumping system, lowering the total life cycle costs.

15.2.3 Parameters to be Checked

– Pump speed and torque control, axial load on drive head thrust bearing

Downhole and surface measuring device and sensors are used to record the axial load with load cells on the PCP drive head thrust bearing, the rod speed and torque and the power supply Variable Frequency Drive to control speed and torque pump motor.

– Pressure inlet and outlet and pump-off control

The lowest possible fluid level back at the PCP inlet allows the highest possible fluid entry into the wellbore from the reservoir's capacity with a maximum oil production. The fluid level over the pump intake is precisely and automatically controlled by downhole and surface sensors such as: casing gas pressure, pump intake suction and discharge pressure to regulate and locate the point of maximum production of any given inflow while protecting the system from a pump-off.

– System protection

Torque limiting protects the motor from excessive torque loads and the rod string against breakage. Stick-slip oscillation damping reduces rod string fatigue failures.

– Energy Savings

Rising energy costs and new environmental protection initiatives drive the artificial lift market to "Intelligent artificial lift solutions". The power flow optimizer concept controls the input power, motor power, rod power, pump power and average efficiency. It ensures better efficiency, reduced electricity costs, energy savings and reduced greenhouse gas emissions for any inflow rate.

The results are increased pump run life and minimized lift costs without compromising on efficiency or reliability.

15.3 MONITORING & PRODUCTION OPTIMIZATION

Downhole and surface monitoring provides real-time information about reservoir dynamics, makes it possible to better understand wells in order to optimize methods and increases production with reduced operating costs per barrel.

15.3.1 Production Monitoring and Real-time Flow Rate Model

Well monitoring and pump controlling provide users with on-time continuous estimated flow and display pump curves such as: rpm, surface tubing pressure, gross liquid rate, surface temperature, tubing pressure, electric parameters, and speed pump with variable frequency power.

Model advanced diagnostics are added to set-up parameters, control and monitor the pumping process. Models are available for motor, head, rod string, pump, flow line, tubing, casing, fluids and reservoir specifications, well completion information.

PCP while running, controllers use data monitoring to adjust all interdependent parameters to optimize and maximize production, improving energy savings and enhancing the reliability of the pumping systems.

15.3.2 Data Communications

SCADA (Supervisory Control and Data Acquisition) is an automated control programmable controllers system with communication capabilities to the well site such as radio, satellite, mobile phone. SCADA transfers data from multi well lift stations to the main office network server, monitoring and controlling system performance.

The data communication systems enable users to access operational data anywhere, anytime to allow faster decision-making to take place so as to safely control and adjust pumping parameters if required.

The ability to quickly assess a situation remotely saves users valuable time in troubleshooting and resolving problems.

15.3.3 Data Capture, Data Logger and Historical Matching Production

From the field information, Data sampler captures real-time information such as pump performance charts, as well as plots of production information. Data logger collects time-stamped faults, warnings, and event logs that can be viewed through the display and network server, such as pump starts, stops, mode changes, electricity used, power up, power loss, overvoltage, over current, pump torque and speed, fluid parameters, customizable alarms, fault history.

The historical matching status of pumps from well(s) can assess a problem in the PCPs and even predict future problems with alarm callouts such as rod breakage and rotor and stator elastomer damage.

15.4 MONITORING EQUIPMENT

15.4.1 Downhole Monitoring

Current downhole monitoring systems which focus on well and pump performance are permanent pressure-temperature gauges to measure pump intake, pump discharge and annulus pressures-temperatures. The ES-PCP downhole monitoring system needs more data such as accelerometers for vibration, voltage and currents.

15.4.1.1 Lufkin – Zenith Oilfield Technology Ltd

The Zenith PCP Protection Z-SIGHT™ Automated well surveillance provides PCP monitoring. It includes downhole rotor position, rod torsion, pump rotation speed and permanent pressure gauge measurements to provide PCP installation knowledge and achieve maximum run life and production.

PCP running, the real time measurement such as rotor depth change can predict an increased load, thermal expansion or drive rod wind up.

Benefits are:

- The rotor position measured downhole improves rotor installation with an optimum position while running.
- The downhole and surface measured rod speeds and the down-hole measured rod torsion control the rod string twist, reduce over torsion and risks of rod breakage.

15.4.1.2 Schlumberger Phoenix Select CTS (Cable to Surface) monitoring tool

The Phoenix CTS system transmits data to the surface *via* an independent encapsulated clad instrument cable. Monitoring parameters include intake pressure, differential pressure, temperature and vibration.

15.4.1.3 GRC Geophysical Research Company Fortress

The Fortress PCP-3500 dual pressure/temperature/vibration sensor is a six-channel sensor measuring intake pressure/temperature, discharge pressure/temperature and X & Y axis vibration. Sensors are located in a carrier run above the pump, data is transferred to the surface by an electric line encased in a stainless steel control line.

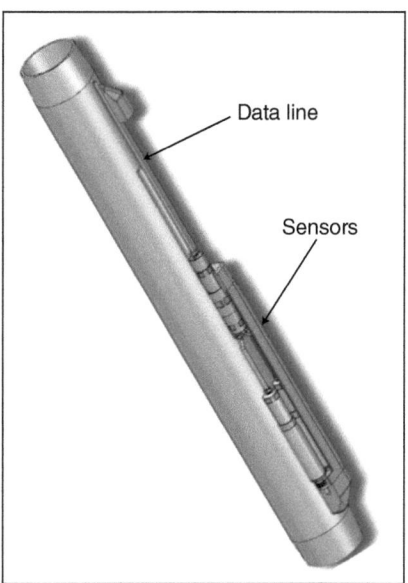

Figure 60
Downhole sensor on tubing.
Source: *GRC*.

15.4.1.4 ZIEBEL Spectra System

The continuous monitoring system records pressure, temperature and up to 3D vibration downhole and surface current frequency spectrum to measure the very nature and shape of the ES-PCP pump motor, vibration and sound of the downhole pump. The system learns the normal sound of your well, takes measurements and notifies you of any on-going changes.

15.4.2 Surface Monitoring

15.4.2.1 Pressure Safety Switch

The simplest is a high pressure safety limit mechanical or electronic type pressure transducer switch connected to a PC controller. The pressure safety switch provides a controlled safety pressure to protect the PCP running from excessive export flow line pressures.

Figure 61

Surface pressure safety switch.

15.4.2.2 The Surface Torque Limiter

The Torque Limiter is one of the first low cost electronic drive head monitors to provide on/off control for the PCP drive head using a torque value as the main control parameter.

15.4.2.3 The Axial Load on the Drive Head Thrust Bearing

The axial load applied by the drive string to the drive head thrust bearing and measured by a load cell is another low cost device. The axial load measurement onto the drive head is related to the well dynamic fluid level.

15.5 WELL MANAGEMENT SOFTWARE

The purpose of the Well Management Software is to use smart downhole and surface measurement well technology to calculate and monitor the dynamic pumping fluid level, to ensure the well is always running at optimum levels and the pump at optimum efficiency in order to optimize production and prevent pump damage.

Data is logged in real time and makes it possible to react specifically to the power controller and the pump by adjusting the pump speed, by monitoring the pump starting torque, pump torque, the back spin torque and by measuring flow rates on view and implement changes remotely at anytime and anywhere.

Downhole and surface sensors used to monitor the PCP are, for example, the axial load measurement on the drive head, the rod torque, the tubing discharge pressure and flow rate, the surface casing pressure, the current and power motor controller, the adjusting speed and torque motor with the variable frequency drive.

15.5.1 Pump Manufacturers Well Management Software

– The PCP Well Manager™ from Kudu – Lufkin Automation

The SAM PCP Well Manager is designed by Kudu and Lufkin Automation. The SAM PCP Well Manager controller works in conjunction with sensors and an electrically powered

Variable Speed Drive (VSD) to optimize fluid production while protecting the pump. The controller measures the pump speed (rpm), the torque and the production rate to establish a speed-torque/production rate relationship. At the point that a step increase in speed does not produce the proportional step increase in fluid production rate, the controller starts to slow the speed by steps until a reduction in fluid production rate is measured. Step by step by repeating the speed increase/decrease sequence the controller evaluates the optimum production rate and the low/high fluid rate limits. Alarms are programmed to prevent pump off and breakage to pumping equipment. The Data Logging also provides real-time management of centralized information using a Supervisory Control and Data Acquisition (SCADA) system. The diagnostic tool also allows a user to remotely control changes.

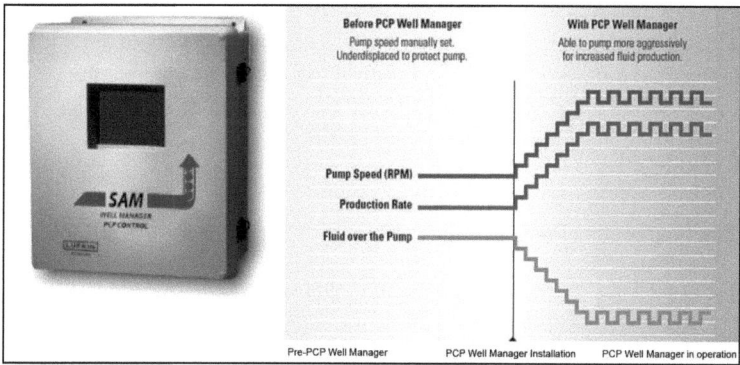

Figure 62-1

SAM™ PCP Well Manager Automation and algorithm.
Source: *Kudu Industries and Lufkin Automation.*

- **The Vector flux drive (VFD) and monitoring read out from Weatherford:**

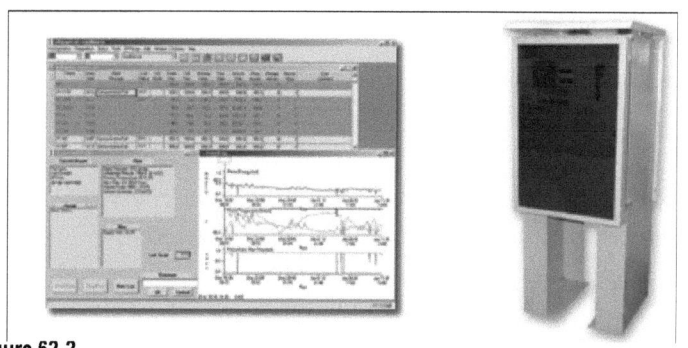

Figure 62-2

Vector flux drive (VFD) and monitoring read out.
Source: *Weatherford.*

- **The ALC 600 PCP and ESP Control from ABB**
- **The Guardian VSD: Variable Speed Drive from R&M Energy Systems**
- **The Monoflo's PCP Controller from Nov**
- **The Lift-Watcher Real-Time Surveillance Service from Schlumberger**

The LiftWatcher provides round-the-clock surveillance, data lift is transferred *via* satellite to and from remote locations, is analysed to identify probable problem causes and quickly reports recommendations and remediation options directly to field personnel to prevent pump problems for rapid implementation.

- **The "PCPOS®" Software from Indrive**

INDRIVE manufactures PCP drive heads and provides software. INDRIVE's "PCPOS®" uses the axial load measurement on the drive head as an expert system to automate in real-time and optimize the PCP pumping system operating at top and control performance.

15.5.2 Well Management Benefits

15.5.2.1 Helps Diagnostics Pump Conditions by

- Receiving data (axial load cell, pressure sensors, Variable Speed Drive (VSD), power, ampere meter, motor speed, etc.).
- Analysing data to make control decisions and monitoring the motor combining precise pump speed/torque control and torque protection.
- Monitoring and controlling fluid level and protecting against pump off.
- Automated operation of diluents or chemical injection system.
- Providing equipment protection for rods, locked rotor, swollen stator elastomer, pump sanded in, rotor re-tag, motor failure, gearbox failure.
- Communicating to remote monitoring by network servers (radio link, cable).
- Sending alarm (s), shut down pump.
- Providing embedded data in the software for rod string, tubing and pump and continuously updating operating history of well(s) and the reservoir.

15.5.2.2 Well Production Benefits

- Efficiently monitor fluid levels.
- Optimize well production.
- Log historical data.
- Analyse well data, performance and make changes remotely.

15.6 MONITORING WELL MANAGEMENT CONCLUSIONS

PCP pumping systems are getting better and smarter with downhole and surface monitoring sensors, with surface controllers and well management software systems and data communication capabilities to the well site, particularly *via* satellite and cell phone.

Downhole monitoring with pressure-temperature-vibration of the pump and the surface monitoring with pressure – temperature – flow rate – torque control – axial load – speed control – Variable Speed Drive (VSD) power unit makes it possible to monitor the artificial lift system, analyse pumping efficiency and adjust pump speed to optimize and increase well production by keeping the fluid at the lowest bottom hole pressure possible.

Furthermore, monitoring detects abnormalities, predicts failures, reduces downtime, work overs and operating costs, minimizes lost production and extends equipment operating life with lowered operating costs per barrel.

Surveillance services are now available to identify real-time probable problems and report recommendations to field personnel so as to prevent damage to the pump.

CHAPTER 16

PCP Software

Artificial lift PCP software is available to design, evaluate performance and select surface drives and bottom hole assemblies progressing cavity pumping applications.

Software is mainly provided by pump manufacturers, pump suppliers and Research Centres to allow end-users to finally select the right pump for the right well.

16.1 C-FER "PC-PUMP®" SOFTWARE

C-FER Technologies based Edmonton Canada operates as a subsidiary of Alberta Innovates – Technology Futures to provide innovative solutions, engineering, testing, and applied research services.

C-FER has developed the PC-PUMP® software, an interactive design and evaluation of a PCP's pumping system www.pc-pump.com/pc-pump-software/design-tools.
 – C-FER provides The PC-PUMP® v3.2.5 software demo version, available on the C FER web site.
 – C-FER performs Course focuses on fundamentals, design, performance optimization and operation of PCP systems and training with the PC-PUMP® software.

The software contains a system configuration module and windows to enter wellbore geometry, casing and tubing sizes information, build a rod string and pump selection to evaluate the downhole lift.

16.1.1 PC-PUMP® Software Modules

16.1.1.1 The Wellbore Geometry Module

It allows an existing wellbore profile to be entered or a new one can be created using the well design mode.

16.1.1.2 The Downhole Equipment Configuration Module and Windows

Wellbore geometry information can be entered, casing sizes, well completion and tubing sizes can be selected, rod strings from various rod manufacturers and of various types can be built, and a pump to evaluate the bottom hole pumping assemblies can be selected.

- The tubular selection window is for selecting casing, tubing, tail joints and diluents injection.
- The rod string selection window is for building a rod drive string with different well characteristics at different depths by using a rod selection database with solid, continuous or hollow rods. Rod guides per rod can be specified or selected by the programme based on a maximum tubing/rod contact force.
- The Pump Selection window is for selecting one or more pumps for analysis from an extensive database containing many pump models.

16.1.1.3 The Surface Drive Equipment Selection

The surface equipment consists of a drive head, electric or hydraulic motors selected from a database of vendor equipment.

In analysis inputs, the fluid properties and operating well conditions are specified.

16.1.1.4 The Single-phase or Multi-phase Fluid Properties

The single-phase (no gas) fluid is specified with actual density, or by the fluid composition allowing the programme to calculate the density.

The multi-phase fluids (solution gas and/or free gas is present) is specified to take into consideration the location of the pump intake (or tail joint or shroud intake if applicable) with respect to the perforations to determine the amount of free gas present in the pump.

An advanced viscosity option is available for single-phase fluid to account for changes in viscosity due to temperature, shear rate, and/or water fraction. For multi-phase fluids the advanced viscosity option allows the dead oil viscosity to be specified as a function of temperature, which will improve the accuracy of the results, especially for heavy oil.

16.1.1.5 Operating Well Conditions

The operating conditions panel is for specifying basic downhole conditions and how the system will be run.

The main operating parameters are: total and vertical depth, a series of flow rate and fluid parameters (total fluid rate and pump speed, the oil rate, the water rate, the free gas rate), the free gas percent, the GOR, the fluid viscosity, the temperature, pressure parameters (fluid level, submergence, the bottom hole pressure).

16.1.1.6 Calculated Variables Include

- Pump intake and discharge pressures.
- Pump pressure loading (as % of rated pressure).

- Hydrostatic head and flow losses.
- Maximum rod torque and axial load.
- Effective rod stress (as % of yield).
- Maximum rod/tubing contact load.
- Energy costs and system efficiency.
- System input power & prime mover output power.

16.1.1.7 Analysis Tools and Displays

The PC-PUMP® evaluates and measures the performance of the downhole system ensuring the pump is adequately sized for pressure and volume and checks rod loads to ensure the rods have sufficient strength for the application, calculates side loading, assesses tubing wear and estimates fatigue life.

16.1.1.8 Energy Flow

Displays efficiency and energy consumed by each component of the system. Also displayed is the breakdown of the pressure, torque and power loads on the pump into wellhead pressure, hydrostatic head, flow losses and friction torque.

16.1.1.9 Drive Equipment

Displays the loading (torque, power, speed, axial load) and other output quantities (e.g. current, motor temperature) relevant to either surface or downhole drive equipment.

16.1.1.10 Basic Fluid Flow

Displays graphs of the casing and tubing pressure profiles, the pressure components (i.e., hydrostatic head vs. flow loss), the flow losses (as caused by the rod body and by any guides or couplings) and the fluid temperature and effective viscosity.

16.1.1.11 Rod Loads and Deflections

Displays in graph form the torque, axial load, effective stress, torque breakdown, deflection and rotation *versus* the measured depth. Also displayed are any space-out requirements from rod stretching due to axial load and thermal expansion.

16.1.1.12 Rod/Tubing Contact

Displays in graph form contact loads, distributed loads, load breakdown, rod/tubing contact locations and rod guide/tubing contact locations *versus* measured depth.

16.1.1.13 Multiphase Flow

Displays in graph form *in situ* flow rates, free and solution gas, flow patterns and volume of pipe occupied by a liquid *versus* measured depth.

16.1.1.14 Key Locations

The key locations table shows the following parameters at important locations in the well.

16.1.1.15 Rod/Tubing Wear

PC-PUMP® can display the rod/tubing wear rates based on the specified system. PC-PUMP® provides the ability to vary the contact materials along the rod string and to display the wear rate as a function of measured depth.

16.1.1.16 Rod Fatigue

The Rod Fatigue module gives the user the ability to model the effects of lower-frequency production fluctuations (e.g. gas or water slugging). A sensitivity analysis is provided for lower frequency fluctuations, since it can be difficult to know how much the loads will vary during slugging.

16.1.1.17 Backspin Analysis

The Backspin module can calculate the response of the system to two major types of rod backspin situations: normal shutdown and seized pump. The brake response can also be calculated if a brake curve is provided by the user.

16.1.1.18 Sand Settling

A sand settling module is available to calculate the minimum flow velocity (setting velocity) to prevent sand from settling back down the well. In most cases, it is recommended that the fluid velocity in the tubing be significantly higher than the settling velocity.

PCP software are now a precious help and useful for the PCP selection and design process, for selecting rotor and stator materials for application compatibility with the downhole fluid, temperature and environment, as well as a suitable rotor-stator fitting for an expected long run life.

CHAPTER 17

Pump Failures & Operating Problems

PCP failures with their respective root causes and possible solutions, failures identification, description, analysis and tracking will reduce premature failures and increase pump run-life.

PCP operating problems identified by analysing sensors and production data will reduce premature troubleshooting, improve lift performances and reduce costs.

Identified pump failures are:

- Rotor failures (wear, heat cracking, fatigue…).
- Stator Failures (fatigue, wear, fluid incompatibility…).
- Rod string failure mechanisms (fatigue, excessive torque…).
- Tubing string failures (wear, corrosion, etc.).
- Well head and pump drive failures.

17.1 PCP FAILURES – IDENTIFICATION AND DESCRIPTION

The progressing cavity pump consists of two parts: The rotor base metal with coating (hard chrome plating) and the stator with three components, a steel tube, a bonding system and an elastomer contour. Under particular circumstances, failures may occur which are related to one of these two components.

17.1.1 Rotor Failures

- Abrasive wear: The rotor wear is caused by excessive abrasive particle concentration in the pumped fluid. Abrasive wear can be just on the surface of the hard chrome plating or down to the base metal and usually located on the crest of the rotor. In either case, the original rotor OD size and profile are changed. Abrasive wear also occurs on the stator elastomer and the rotor-stator wear reduces the rotor-stator interference seal fit to become a clearance fit. Wears allow slippage or recirculation fluids back to the suction through this clearance. The wear gradually affects the pump's performance and flow rate will gradually decrease.

- Metallic parts from the well completion or milling displaced by the fluid to the inlet of the pump can be lodged in cavities causing extensive damage to rotor and stator.
- Acid attack occurs when the pH of the produced fluid drops below 6.0 resulting in a stripping of the chrome plating on the rotor. This usually occurs with an acid well stimulation while leaving the pump in the well.
- Fatigue failure is the result of the rotor material undergoing cyclic stresses.
- Torsion fatigue can occur if the pump rotor locks up for any reason such as solids building up or elastomer failure. The upper portion (rotor and rod drive) will continue to rotate and twist off until it breaks down if it is not immediately stopped on the surface.

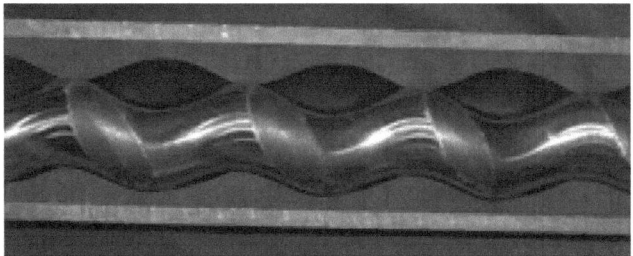

Figure 63-1

The crest of the rotor wear.
Source: *C-Fer*.

17.1.2 Stator Elastomer Failures

- Pump run dry: A pump running dry cause rapid excessive heat up. Normally the presence of liquid in the pump provides a lower friction sliding contact and a cool down of the elastomer. In the absence of liquid, the rotor runs dry on the stator causing elastomer overheating. The elastomer becomes hard, brittle and cracked will fail quickly. If the surface sensors detect an increase in torque and power drive requirements, the pump must be stopped.
- Elastomer chemical and thermal swell:
  - The chemical swell is the dissolution of fluids, liquid and gas through the elastomer surface then diffusion in the elastomer matrix. The incompatibilities between the stator elastomer and the produced fluids (production fluids and/or treating chemicals fluids) at downhole temperature can cause elastomer chemical attacks resulting an elastomer expansion. The chemical swell is generally caused by the presence of aromatics, benzene and other compounds in the pumped fluids. The chemical swell is generally permanent and non-reversible.
  - The thermal swell is a thermal elastomer expansion due only to an elevation in temperature. The thermal swell can be predicted and is usually not permanent. It decreases with temperature.

  The main consequences of elastomer swelling (chemical or thermal) are the expansion volume resulting in an increase in friction interference between rotor/stator and a sta-

tor over-heat will cause deterioration of both stator and rotor and on the surface higher torque and power requirements. If surface sensors detect an increase in torque and power drive requirements, the pump must be stopped.
- Gas permeation / explosive decompression: gas permeation occurs when free gas, under downhole conditions (pressure, temperature and time) enters the elastomer matrix. The gas will expand when the pressure is reduced. For example, when the stator is retrieved to the surface, the gas expands inside the elastomer matrix; it often results in blisters or bubbles forming within the elastomer matrix. It can sometimes expand to the point that the elastomer ruptures. This is referred to as "explosive decompression".
- Bond failure: A bond failure occurs when the fluids penetrate through the interface at the stator ends, dissolving the bonding agent between the elastomeric lining and the stator tube.

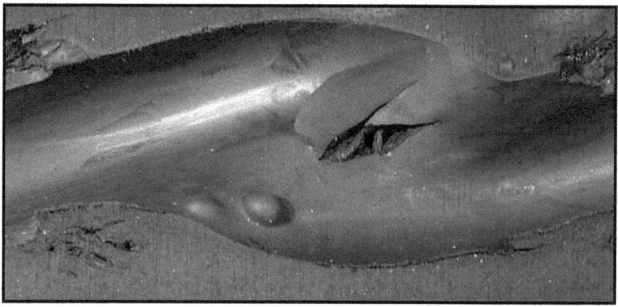

Figure 63-2

High GOR producer elastomer bubbles and rupture.
Source: *Baker*.

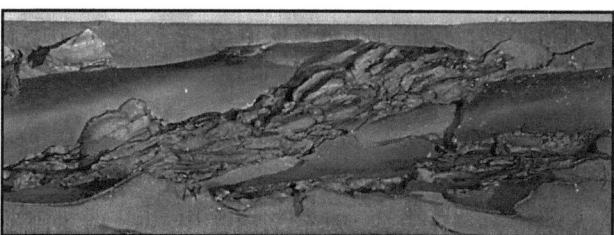

Figure 63-3

Elastomer damaged by high discharge pressure and heat.
Source: *Baker*.

Recommendations: To predict and reduce elastomer failures, manufacturers recommend elastomer sample tests in a laboratory to define the exact elastomer for a specific downhole fluids conditions and location. Tests can predict how elastomers react to specific well fluid on downhole conditions and permit prediction of elastomer dimensional change, mainly due to fluid composition, thermal expansion and chemical swell.

17.2 OPERATING PROBLEMS

From the well: No flow = the fluid doesn't reach the well head.

When downhole PCP or surface problems arise such as NO flow, less flow, loss of suction or excessive power consumption. The pump has to be stopped and causes have to be identified.

17.2.1 No Flow Upon Start-up

Check:
- The export flow line and valves positions, if it is closed?
- At the start, the correct direction rotor rotation to eliminate risks to unscrew one rod joint.
- The power drive current intensity, is it lower than normal.
- Excessive discharge pressure.
- Improper positioning of the rotor level inside the stator. Rotor above stator.
- Breaking down of the string driving the rotor.
- Pump wear.

17.2.2 Less Flow

Less flow can be caused by:
- The rotor is positioned too high in the stator. Results will be reduced pump efficiency, induced vibration, damaged upper stator and rotor fatigue.
- An internal recirculation flow (slipping flow, slippage or leakage) due to:
  - an increase of clearances between rotor and stator, (rotor-stator wear),
  - an excessive discharge pressure.
- The slipping flow also depends on the fluid viscosity related to the temperature.
- A loss of suction, the fluid is not reaching the inlet PCP at sufficient pressure to maintain pump efficiency. The loss of suction can be caused by the pump being unable to prime, or cavitation by insufficient inlet pressure or gas content.

17.2.3 Excessive Power Consumption, Higher Torque

PCPs are volumetric and the power (torque and speed) to drive solid rods and rotor is nearly proportional to differential pressure (torque), to volume (speed) at a fluid viscosity. If either increase, the required input power will also increase – higher operating pressures require more power.

If the pressure and speed parameters are not changed with the power drive pump increase, pump failure can occur.

The current intensity is higher than normal:
- The rotor is installed too deeply on the tag bar, drive string in compression, high friction rotation with tag bar and heat generated.
- Sand building up above the rotor.
- Cuttings sucked up by the pump.
- Excessive swelling of the stator due to chemical agents.
- The "Pump-off" is not sufficiently liquid at the pump inlet. The pump running dry causes a lack of lubrication to the stator, the torque increase resulting in extremely high temperatures being generated. The high temperature ultimately burns the elastomer. "Pump-off" may be caused by one or a combination of the following problems: plugged pump intake, poor inflow, or production rates exceeding inflow.
- Excessive slippage and flow back to suction due to excessive discharge pressure.
- Accidental shut-off of a surface valve upon discharge.

17.3 SURFACE MECHANICAL PROBLEMS

17.3.1 Stuffing Box

Excessive stuffing box leakage may be due to:
- Lack of adjustment of the packing elements.
- Wear of the packing elements.
- Wear of the drive shaft (case of wells containing sand).

Check lubrication and joints to extend the life of the shaft and avoid excessive leakage. However, slight leakage helps to disperse excessive heat in the stuffing box and lubricate it.

Consequently, the stuffing box should be adjusted properly so that it is:
- Not too tight, so as to avoid premature wear of the packing and of the shaft, and/or overloading of the motor.
- Not too loose, so as to avoid excessive leakage.

17.3.2 Bearing on Drive Shaft

Improper lubrication may result in:
- Excessive heat and noise generally indicates premature failure of the bearings.

17.3.3 Vibration Drive Head

The possible causes of vibration are:
- Bearing failure.
- Bending of the rod or polished rod.
- Drive string harmonics.
- Other problems with the downhole equipment.

CHAPTER 18

PCP Advantages and Limitations

The Progressing Cavity Pump is made up of a helical steel rotor which rotates inside a double internal helical stator.

The Progressing Cavity Pump is a volumetric displacement pump, the pumped flow rate is controlled by rotor rotation, the rotor being coupled to the rod string and to the surface drive head. The ES-PCP rotor is driven by a bottom hole ES electric motor.

18.1 PCP ARTIFICIAL LIFT ADVANTAGES

PCP pumping systems have many advantages in terms of concept, characteristics, equipment, operating conditions, maintenance, investments and operating costs compared to other traditional artificial oil lift technology.

PCP's advantages compared with other artificial lift methods are summarised as follow:
- Used with various reservoir types and capable of handling a wide range of fluids from water and light oils to highly abrasive heavy oils.
- Especially designed for heavy crude oil and high sand-cut reservoir, diluted heavy oil, thermal recovery as well as gas well dewatering.
- Manufacturers provide a wide range of PCP models for various flow rates up to 2000 m^3/day maximum, pumping depths of up to 1800 m (6000 ft) and fluid well conditions.
- PCPs have higher pump efficiency (55%-75%), much higher than artificial lift methods such as Electric Submersible Pumps or Beam Pumps.
- This results in energy savings, lower running costs, lower initial investment costs, the lowest costs per barrel of recovered fluids.
- Adjustable rotor rotation from 25 to 500 rpm with a variable speed drive (VSD) power and controls drive head motor. The pump speed rotation is mainly proportional to pumped flow rate; it is used to adjust production with the reservoir capacities and fluid characteristics. The PCP's pumped flow rate is not directly a function of the gain of pressure, even though it affects pump efficiency.
- The PCP's volumetric displacement is continuous with no impulse pressure and volume during the pumping process. The absence of pressure pulsations in the formation

near the well generates less sand production in unconsolidated reservoirs. The constant flow production facilitates the instrumentation. This is not the case with beam pumps.
- The PCP's volumetric displacement provides low internal shear rates, less fluid emulsification to be caused compared to ESP centrifugal pump. The oil and water separation on the surface will be much easier.
- The PCP tolerates high percentages of free gas with inlet gas separator. Free gas does not block the pump as centrifugal pump but lowers the pump capacity and reduces the efficiency as with any type of standard pump.

18.2 PCP'S ACTUAL LIMITATIONS

PCPs also have several technical limitations, summarised as follow:
- Limited temperature capability (maximum of 180°C or 350°F) with stator elastomer.
- Require a better selection and PCP design process with rotor and stator materials, as well as a suitable rotor-stator fitting for an expected long run life.
- The PCP pump combination of a high volume and high head capacities is not possible compared to ESP centrifugal pump.
- Stator elastomer key of the PCP's stator depends on the fluid environment characteristics at downhole conditions. Manufacturers provide a large variety of elastomers and make laboratory tests compatible to define the exact elastomer compounds for a specific application.
- The rod string rotation provides rod/tubing wear in the deviated section. Rod centralizers need to be well positioned on specific deviated sections to reduce the tubing wear.

18.3 PCP EQUIPMENT ADVANTAGES

PCP Surface and Downhole Equipment

The drive head is simple and compact, light, can be installed easily. Less space is needed to install a PCP on a multiple well cluster, especially suitable for land and offshore applications. Low maintenance: only the sealing and bearings of drive head need to be maintained regularly.
- PCP downhole equipment. To install the PCP stator with tubing, the rotor with the rods string are limited works using lift or work over rig and the same for servicing the pump.
- Low energy consumption and lower operating costs because as a result of high pump efficiency.
- Low noise levels while drive head running.
- Drive head is compact with a limited visual impact, possible uses with urban areas.
- Can be used on offshore production platform for limited uses.

- PCP driven with new permanent magnet motors (PM) as described in chapters 9-4.3 on surface and 12-3 on downhole. Permanent magnet motors have recently been marketed on the surface to direct drive rods string and downhole as semisubmersible motors. To summarise the advantages of PM motors and specific benefits compared with standard asynchronous motors:
  • Higher efficiency with lower heat production compared to asynchronous motors and capabilities of a wide variable speed range, high torque at low speed, the torque decreases as speed increases.
  • PM magnet technology decreases the motor dimensions, eliminates complex satellite gear box, the weakest and most costly component on standard ESP-PCP.
  • The surface Variable Speed Drive provides maximum speed and torque control capabilities with smooth motor start and shut down for operating wells with unstable production.
  • An increase in global production with an extended run life enhances the reliability of artificial lift pumping systems and brings about substantial operating and capital cost savings.

18.4 PCP WELL MONITORING AND DATA TRANSMISSION

PCP pumping systems getting better and smarter with downhole monitoring sensors such as pressure-temperature-vibration of the pump and surface sensors, pressure-temperature-flow rate-torque control-axial load-speed controled by Variable Speed Drive (VSD) power unit.

Monitoring makes it possible to closely control the artificial lift system by using surface controllers, well management software system by analysing pump efficiency and adjusting pump speed to optimize and increase well production by keeping the fluid at the lowest bottom hole pressure possible.

Furthermore, monitoring detects abnormalities, predicts failures, reduces downtime, walkovers and operating costs, minimizes lost production and extends equipment operating life with lower operating costs per barrel.

18.5 PCP WELL MANAGEMENT AND SOFTWARE

The data communication systems enable users to access operational data anywhere and at any time, meaning faster decision-making in order to safely control and adjust pumping parameters if required.

Well management surveillance services identify real-time probable problems and report recommendations to field personnel.

PCP software helps and provides assistance for better PCP rotor and stator selection, material selection compatibilities with downhole fluid, temperature and the environment and to design the pumping process.

CHAPTER 19

Units, Pump Parameters, Nomenclature

19.1 CONVERSION FACTORS

Length: Meters (m), Feet (ft), Inches (in)
1 meter = 3.28 ft = 39.6 inch
1 in = 0.0254 m = 25.4 mm
1 ft = 12 in = 0.3048 m

Area: square meters (m^2), square feet (ft^2)
1 square feet = 0.0929 m^2
1 square meter = 10.76 Sq feet

Volume and Flow rate: Litre (l), cubic metre (m^3), cubic feet (ft^3), Barrel (bbl), US gallon (gal US or USgal), Imperial gallon (gal) (UK)
1 m^3 = 35.31 cubic feet
1 cubic feet = 0.02832 m^3
1 m^3 = 6.28 bbl
1 bbl = 158.9 liters = 0.1589 m^3 = 42 USgal = 34.97 Imperial Gallons (UK)
1 USgal = 3.785 liters = 0.0238 bbl
1 gal (UK) = 4.545 liters = 0.02859 bbl = 1.200 gal US

Flow rate: cubic metre/day (m^3/d), m^3/h, barrel/day (bbl/d), US gal/minute (USGPM)
1 USgal/minute = 34.3 bbl/d = 5.451 m^3/d
1 bbl/d = 0.0662 m^3/h

Masse: kilogram (kg), Pound (lb)
1 kg = 2.204 lb
1 lb = 0.4536 kg

Force: kilograms-force (kgf), pound force (lbf).

The SI force units is the Newton (N), 1 daN = 10 N

1 kgf = 9.81 N = 2.204 lbf

1 daN = 2.248 lbf

1 lbf = 4.448 N = 0.4448 daN = 0.4536 kgf

Torque: Newton-m (N.m), DaN.m, kilograms-force-metre (kgf.m) or meter-kilograms force, pound-force-feet (lbf.ft) or foot-pounds-force (ft.lbf)

A "moment or bending moment" is the general term used for the tendency of an applied force to rotate an object about an axis with no change to the angular moment of the object. If the angular moment of the object is changed it is a "torque". The "torque" is the tendency of an applied force to rotate an object about an axis and to change the angular moment.

1 daN.m = 7.37 ft.lbf

1 ft.lbf = 0.13825 kgf.m = 0.1356 daN.m = 1.356 N.m

1 kgf.m = 7.23 ft.lbf

Note: the Surface tension unit is the kilograms-force per meter or pounds-force per feet 1 lbf/ft = 1.488 kgf/m = 1.46 daN/m.

Pressure: Pascal (Pa) = one Newton force per square metre (N/m^2), kilopascal (kPa), mega Pascal (MPa), Bar (bar), Pounds per square inch (psi):

One kilopascal (kPa) is about 1% of atmospheric pressure (near sea level) (atm).

1 Pa = 1 N/m^2 = 10^{-5} bar = 1.019 kgf/cm^2 ≈ 1 kgf/cm^2 = $1,42 \cdot 10^{-4}$ psi

1 kPa = 0.145 psi

1 MPa = 0.145 kpsi

1 MPa = 10^6 Pa = 10^6 N/m^2 = 1 N/mm^2 = 100 N/cm^2 = 10 bar

1 bar = 100 kpa = 10^5 Pa = 0.1 Mpa = 0.987 atm = 14.5038 psi

1 kgf/cm^2 = 9.81 × 10^4 N/m^2 = 9.81 × 10^4 Pa = 0.981 bar

1 psi = 0.0689 bar = 6.895 kPa

1 kpsi = 6.895 MPa

The pressure in meters of water column (mwc), 10 meters of H_2O = 1 kgf/cm^2 = 9.81 N/cm^2 ≈ 1 bar

Fresh water gradient = 0.433 psi/ft = 9.8 kPa/meter

Temperature: degrees Celsius (°C) or Fahrenheit (°F), Kelvin (°K).

t°F = 9/5(t°C + 32), t°C = 5/9(t°F − 32)

t°Kelvin = t°C + 273

T°C	T°F	T°C	T°F
0	32	204	400
104	200	260	500
148.9	300	316	600

Yield strength

The yield strength or ultimate strength is the elastic limit before permanent deformation.
Strength unit: the MPa = $1 N/mm^2$ = 1000 kPa = 145 psi

Dynamic viscosity μ

SI dynamic viscosity unit is the Pascal.second (Pa.s)
CGS viscosity unit is the poise (cP)
1 cP = 10^{-3} Pa.s = 0.001 Pa.s
1 Poise = 0.1 Pa.s
Water at 20°C has a viscosity of 1.0020 cP.

Kinematic viscosity $v = \dfrac{\mu}{\rho}$

ρ density SI unit kg/m^3
v kinematic viscosity SI unit is the m^2/sec
v kinematic viscosity CGS units is the Stokes (St), 1 cSt = 1 mm^2/s.
The kinematic viscosity of the water = 1 cSt at 20°C

Power: The SI power units is the Watt (W), Horse power (hp)
$power(W) = torque(N.m) \times angular\ speed = torque(N.m) \times 2\pi \times rotational\ speed(rps)$

$$power(kW) = \frac{torque(N.m) \times 2\pi \times rotationnal\ speed(rpm)}{60,000}$$

$$power(hp) = \frac{torque(lbf.ft) \times 2\pi \times rotationnal\ speed(rpm)}{33,000}$$

1 KW = 1 KJoule/sec
1 hp = 736 W
hp = rotational speed (rpm) × torque(ftlbs)/5252

API gravity = (141.5/Specific gravity) − 131.5
Specific Gravity = 141.5/(API Gravity + 131.5)
fresh water specific gravity = 1.0

19.2 PCP PARAMETERS

Multi-lobe pump ($L_r - L_s$)
O_1 centreline of the stator axis
O_2 centreline of the rotor axis

$E = O_1O_2$ distance = rotor/stator eccentricity or eccentricity of the pump

L_r number of rotor lobes

L_s number of stator lobes $L_s = L_r + 1$

       multi-lobe $(L_r - L_s)$ pump example: 1-2, 2-3, 5-6

P pitch length = length of 360° rotation of the crest trace of one of the helix rotor or stator lobes

P_r rotor pitch length (meter or in.) $P_r = \dfrac{L_r}{L_r + 1} \times P_s$

P stator pitch length of the stator (meter or in) $P_s = \dfrac{L_r + 1}{L_r} \times P_r$

C number of stator cavities $C = L_r \left(\dfrac{H_s}{P_r} - 1 \right)$

i pump kinematics ratio $i = L_r / L_s = P_r / P_s$

Mono-lobe 1-2 pump

D thickness of a single helical rotor (minor rotor diameter)

$D + 2E$ diameter of the helix rotor (major rotor diameter)

D minor diameter of the helix stator (minor stator diameter)

$D + 4E$ major diameter of the helix stator (major stator diameter)

$4E \times D$ = Flow area cross section of the pump = the area of fluid is constant

L_r number of rotor lobes = 1

L_s number of stator lobes = $L_r + 1 = 2$

P_r pitch length of the rotor = 1

P_s pitch length of the stator = the pitch stator length of a cavity

$$P_s = \dfrac{L_r + 1}{L_r} \times P_r = 2\,Pr \qquad P_r = P_s/2$$

$$P_s \times L_r = P_r \times L_s$$

i pump kinematics ratio $i = L_r / L_s = P_r / P_s = 2$

H_s length of the stator

C number of effective enclosed cavities $C = \dfrac{H_s}{P_s} - 1 = \dfrac{H_s}{2P_r} - 1$

V pump cylinder capacity = Flow area $\times P_s$ = volume of one stator pitch length P_s

$V = 4E \times D \times P_s$ = cavity volume for one rotation

N_r number of rotor rotations per minute (rpm)

Q flow rate (l/mn or gal/min = gpm)

Q_c calculated pump flow rate per minute (non slippage) $Q_c = 4E \times D \times P_s \times N$

Q_s pump leak (slip, slippage or leakage)

Q_a actual flow rate or total flow at operating load determined by considering a leak rate Q_s

$$Q_a = Q_c - Q_s$$

δp pressure reference for a cavity

n_p number of stator pitches of length P_s

n_r number of rotor pitches of length P_r

ΔP total pressure head rating $\Delta P = \delta p(2n_p - 1)$

ρ pump efficiency (for the evaluation $\rho = 0.7$).

$V(\text{cm}^3)$ pump cylinder capacity

$\Gamma_{(\text{daNm})}$ maximum resistant torque (in daN-m or ft-lbs)

$$\Gamma_{(\text{daNm})} = 1.63 \times V_{(\text{cm}^3)} \times \Delta P_{(\text{kPa})} \times 10^{-5} \times \rho^{-1}$$

μ_f viscosity of the effluent at the inlet temperature (in cP)

μ_s viscosity of the effluent at the surface (in cP)

L length of the tubing (in m)

T temperature at pump inlet

t°C   temperature in Celsius (°C)

t°F   temperature in Fahrenheit (°F)

mwc or mH$_2$O meters of water column (meters of H$_2$O)

19.3 PCP's USEFUL FORMULAS

Total Net Lift (mwc or ftwc) = Pumping Fluid level (m or ft) x fluid specific gravity (SG) + Flow line back head pressure (mwc or ftwc) + Head Loss or friction loss of fluid up the tubing (mwc or ftwc).

Fresh water specific gravity = 1 at 4°C (39°F)

Fresh water gradient = 9.8 kPa/metre = 1.421 psi/m = 0.433 psi/ft

From Water Pressure to Head Water column in metre (mwc)

$$\text{Head Pressure (mwc)} = \frac{\text{Pressure (kPa)}}{9.8} = \text{Pressure (kPa)} \times 0.102 = \text{Pressure (bar)} \times 10.2$$

$$\text{Head Pressure (mwc)} = \frac{\text{Pressure (psi)}}{1.42} = \text{Pressure (psi)} \times 0.704$$

- One metre of water column (mwc) = 3.281 ftwc = 9.8 kPa = 1.42 psi
- 10.2 metres of water column (mwc) = 33.46 ftwc = 100 kPa = 1 bar = 14.5 psi

From Water Pressure to Head Water column in foot (ftwc)

$$\text{Head Pressure (ftwc)} = \frac{\text{Pressure (kPa)}}{2.98} = \text{Pressure (kPa)} \times 0.3347 = \text{Pressure (bar)} \times 33.47$$

$$\text{Head Pressure (ftwc)} = \frac{\text{Pressure (psi)}}{0.432} = \text{Pressure (psi)} \times 2.32$$

- One foot of water column (ftwc) = 0.304 mwc = 0.432psi = 2.98kPa = 0.0298bar

From Fluid Pressure to Head Water column in metre (mwc)

Fluid Specific gravity = SG

Head Pressure (mwc) = Pressure (kPa) × 0.102 × SG = Pressure (bar) × 10.2 × SG

Head Pressure (mwc) = Pressure (psi) × 0.704 × SG

From Fluid Pressure to Head Water column in foot (ftwc)

Head Pressure (ftwc) = Pressure (kPa) × 0.3347 × SG = Pressure (bar) × 33.47 × SG

Head Pressure (ftwc) = Pressure (psi) × 2.32 × SG

Power P in kW or HP horse power

$$P(\text{kW}) = 1.046 \times 10^{-3} \times \Gamma(\text{daN.m}) \times N(\text{rpm})$$

$$P(\text{HP}) = \frac{\Gamma(\text{ft.lbs.}) \times N(\text{rpm})}{5252}$$

Torque in daNm = (Total Net Lift(m) × Nominal Displacement (m³/day/100 rpm)/921 + Friction torque (daNm)

Torque in ft.lbs = (Total Net Lift (m) × Nominal Displacement (m³/day/100 rpm)/125 + Friction torque (ft.lbs)

Solid Rod Speed (rpm) = Production Rate (m³/day)/Pump Nominal Displacement (m³/day/rpm)

19.4 NOMENCLATURE

Symbol	Definition	Unit
°API	Degree API	dimensionless
SG	Specific Gravity or Relative Density	dimensionless ratio liquid density to water reference at 4°C, SG water = 1
Yg	Gas specific gravity	ratio density of gas to air reference at T°C, SG air = 1 at NPT
Yo	Oil specific gravity	ratio density of oil to water reference at T°C, SG water = 1 at NPT
μ	Dynamic Viscosity (Poises)	cP, centipoises = 1 mPa.s
ν	Kinematic Viscosity (Stokes)	cSt, centistokes = 1 mm^2.s^{-1}
ρ	Density	mass per unit volume
	Density of water	1 at 4°C
	Density of dry air at NPT	1,025 kg/m^3 at 15°C
K	Permeability (Darcy)	mD = 10^{-3} D
bbl	Volume	m3, bbl (1 barrel = 159 litres)
NPT	Normal Temperature and Pressure	at 20°C (293,15K, 68°F) and 1 atm (101,325 kN/m^2, 101,325 kPa, 1,013 bar, 14,7 psi)
IP or PI	Productivity Index	Volume/Pressure m^3/d/bar or bopd/psi
GLR	Gas Liquid ratio	%
GOR	Gas Oil ratio	%
B$_o$	Formation volume factor	m^3/m^3
B$_g$	Gas volume factor	m^3/m^3
Q	Flow rate, Volume/time unit	m^3/d or bbl/d – bopd
P	Pressure	kPa, MPa, bar, psi
Pb	Bubble point pressure	kPa or psi
Z	Gas compressibility factor	0.81 to 0.91
T	Temperature	°C or °F
	Absolute temperature	Kelvin °K = °C + 273
D	Inside diameter of tubing	
d	Rod diameter	
TS	Tensile Strength, Yield strength	MPa, psi
	Ultimate strength	MPa, psi elastic limit
Γ	Torque	daN.m, kg.m or ft.lbf or ft-lbf
F	Load force	daN, kN, lb or lbf or lbf (pound force)
N	Rotation speed	rotation per time (rpm per minute)
P	Power	kW, hp
	Frequency	Hz, Hertz
	Pump efficiency	dimensionless

CHAPTER 20

Web Sites and Manufacturers

MANUFACTURERS & SUPPLIERS		PCP rotor stator	Rod drive	Downhole equip	Drive head And parts	ESPCP	Variable Frequency drive	Surface Equip & Monitoring
ADVANTAGES Products	www.advantageproductsinc.com			X	X			
ALBERTA OIL Tool Dover	www.albertaoiltool.com							
APEX Advanced Solutions	www.apex-advanced.com				X	X	X	
Artificial Lift Co	www.alcesp.com							
BAKER	www.bakerhughes.com				X	X		
BIW ITTCannon Well head feed thru	www.ittcannon.com/BIW							
BORETS part of Weatherford	www.borets.com/pumps.php					X		
BossCLAMP cable protector	www.boss-clamp.com			X				
C-FER	www.cfertech.com							
CALDYNE-TORUS	www.caledyne.co.uk			X				
CAMERON CAMROD, CAM-PC	www.c-a-m.com	X	X	X	X			
CANAM Pipe & Supply Twister	www.canamservices.com	X	X	X	X			X
CAN-K	www.can-k.com							X
CANLIFT Bulgaria	www.canlift.eu						X	
CHINA-OGPE manufacturers &suppliers	www.china-ogpe.com		X			X		
CHRISCOR	www.chriscortools.com							
CJS Coiled Tubing	www.cjsflatpak.ca		X					
COTECO	www.coteco.ca	X	X	X				
Daqing Jinghong	www.jh-pm.com			X				

MANUFACTURERS & SUPPLIERS		PCP rotor stator	Rod drive	Downhole equip	Drive head And parts	ESPCP	Variable Frequency drive	Surface Equip & Monitoring
Direct Drive PCP	www.directdrivehead.com				X			
Dura Product	www.duraproductsinc.com				X			
Dyna-Drill/Lift	www.dyna-drill.com	X						
ENGT Energy & Technology Corp	www.engt.com							
ESO Lift	www.esplift.com							
ESP pump The ES story	www.esppump.com							
EUROPUMP	www.europump.ca	X			X			
Evolution Oil Tools	www.eotools.com							
FHE Frank Henri Equipment	www.frank-henry.com			X				
GRC	www.grcamerada.com							
GRENCo	www.grenco.com	X	X		X			
HALLIBURTON	www.halliburton.com			X				
KACHELLE	www.w-kaechele.de	X						
KUDU	www.kudupump.com	X		X	X			X
MasterOilTool	www.masteroiltool.com							
LUFKIN ZENITH	www.lufkin.com www.zenithoilfield.com			X				X
NEPTUN	www.neptun-gears.ro			X	X			
NETZSCH	www.netzsch.ca www.netzschusa.com							
Millennium Oilflow	www.mostoil.com			X	X			
National Oilwell Varco NOV Monoflo Pacific Artificial Lift Wilson Artificial Lift	www.nov.com www.monoflo.com www.ppptl.co.nz www.iwilson.com	X	X	X	X			
NOVOMET ES PM motor	www.novomet.ru/eng/					X	X	
NORRIS (Dover Co)	www.norrisrods.com							
OLILIFT (Dover Co)	www.oillifttechnology.com	X	X	X	X			
PCP International Supplier	www.pcpinternational.com							
PCM pompe moineau	www.pcmpompes.com	X			X			
PC-PUMP	www.pc-pump.com							
PCP Oil Tools	www.pcpoiltools.com			X				
PEAK Completion	www.peakcompletions.com			X				
PREMIUM ALS	www.premiumALS.com		X					
PROTEX	www.protexcis.com	X			X			

MANUFACTURERS & SUPPLIERS		PCP rotor stator	Rod drive	Downhole equip	Drive head And parts	ESPCP	Variable Frequency drive	Surface Equip & Monitoring
PUMP ZONE	www.pump-zone.com							
ROGTEC	www.rogtecmagazine.com							
R&M Energy Systems	www.rmenergy.com	X		X	X			
Rod Guide Industries Rode drive guide	www.rodguideindustries.com							
SCOPE Product	www.scopeproduction.com							
SCHLUMBERGER	www.slb.com							X
SHENGLI Elec Co	www.e-risun.com				X			
SUN Engineering Inc Cross coupling cable protector	www.sunengineeringinc.com							
TANROC	www.tanroc.com	X		X	X			
TENNARIS rods, hollow rod	www.tenaris.com		X					
TIERRA Alta(CA)	Alberta (CA)	X						
True North Supply	www.truenorthsupplyltd.com	X	X		X	X		
WEATHERFORD COROD TRICO BMW EVI oil tools GEREMIA	www.weatherford.com	X X	 X X X		X			
WOOD Group G Elect ESP pump	www.woodgroup-esp.com/pages/home.aspx							
ZEITECS sensors	www.zeitecs.com			X				

CHAPTER 21

PCP Normalization ISO 15136 -1:2001, -2:2006, -1:2010

> The following documents are extracts from the PCP ISO 1513-1:2001(E)
> © ISO 2001. All rights reserved.

ISO (the International Organization for Standardization) is a worldwide federation standards bodies.

Introduction

"This part of ISO 15136 has been developed by users/purchasers and suppliers/manufacturers of progressing cavity pumps (PCP) for artificial lift use in the petroleum and natural gas industries worldwide. This part of ISO 15136 is intended to give requirements and information to both parties in the selection, manufacture, testing and use of progressing cavity pumps. Further this part of ISO 15136 addresses supplier/manufacturer requirements, which set the minimum parameters with suppliers/manufacturers must comply to claim conformity with this part of ISO 15136."

"A progressing cavity pump comprises two helical gears, one rotating inside the other. The stator and rotor axes are parallel and spaced between each other. The external helical gear (stator) has one more thread (or tooth) than the internal helical gear (rotor). Whatever the number of threads of the two elements, they must always differ by one. The fluid moves from suction to discharge. The discharge and the suction are always isolated from each other by a constant length seal line."

21.1 PCP ISO NORMALIZATION DATED 2001-2006-2010

ISO 15136 -1:2001 for *"Downhole equipment for petroleum and natural gas industries – progressive cavity pumps system for artificial lift* are under the norm".

- annex A (normative) *"Example of performance curves for pump selection"*
- annex B (normative) *"PCP test data sheet"*
- annex C (informative) *"Application design specification data sheet"*
- annex E (informative) *"Engineering methodology"*
- annex F (informative) *"Description of PCP system"*

ISO 15136 -2:2006

The ISO 15136-2:2006. Provides requirements and information on progressing cavity pump surface-drive systems.

- annex A (normative) *"Requirements for elastomers and non-metallic materials"*
- annex B (informative) *"Braque system evaluation method"*
- annex C (informative) *"Installation guidelines"*
- annex D (informative) *"Operation guidelines"*

ISO 15136 -1:2010-04

The ISO 15136-1:2010-04. Provides grades requirements for design validation, quality control and functional evaluation allowing the user/purchaser to select each for a specific application.

- annex A (normative) *"Requirements for progressing cavity pump elastomers"*
- annex B (normative) *"Design validation"*
- annex C (normative) *"Function evaluation"*
- annex D (informative) *"Optional information for PCP elastomer testing and selection"*
- annex E (informative) *"Installation guidelines"*
- annex F (informative) *"Operation guidelines"*
- annex g (informative) *"Supplemental information for PCP performance characteristics"*
- annex H (informative) *"Example user/purchaser PCP functional specification form"*
- annex I (informative) *"Analysis after use"*
- annex J (informative) *"Selection and use of drive-string equipment in PCP applications"*
- annex k (informative) *"Repair and reconditioning"*

For the PCP ISO 15136 -1:2001

Part 1; Pumps norm define:

1 Scope

2 Terms and definitions

3 Symbols

4 Functional specifications

5 Technical specifications

5.4 Design verification

- *flow capacity of PCP considering the rotor and stator size*
- *the head capacity of the PCP considering the number of stator cavities*
- *the tightening between rotor and stator versus temperature*

5.5 Design validation

5.6.1 Validation parameters

To verify flow and head capacity of each pump, the following test shall be conducted with water at a rotating speed of 500 rpm, excepted for high volume/high pressure pumps where speed can be reduced to limit power consumption as agreed between supplier/manufacturer and user. /purchaser:

- *at zero differential pressure and zero leakage:*
- *at maximum operating differential pressure at a target of 15% leakage (minimum 10%, maximum 20%)*

The resulting performance curve is the performance baseline for acceptance testing of the PCP. Consideration shall be given to a swelling test on an elastomer sample with the anticipated crude or equivalent.

5.5.2.2 Calibration

6 Supplier/manufacturer requirements

6.2.2 Product identification

6.3.1 Stator code

21.1.1 Stator Code

Each stator shall have the following code permanently impressed on the exterior

vvv/hh/eee

*vvv is the flow rate, (volume of fluid pumped per time unit) expressed in cubic meters per day at **500** rpm and zero discharge pressure,*

hh *is the maximum head rating (maximum allowable differential pressure) of the PCP: expressed in megapascals (MPa)*

eee *is the manufacturer's code for the elastomer,*

21.1.2 Rotor Code

Each rotor shall have be permanently impressed with the following code:

vvv/hh

vvv is the flow rate, expressed in cubic meters per day at **500** rpm and zero discharge pressure,

hh is the maximum head rating of the PCP: expressed in megapascals (MPa).

Annex A
(normative)
Example of performance curves for pump selection

EXAMPLE

Rotor code	3/12
Temperature	50 °C
Outside diameter of stator	85 mm
Outside diameter of rotor	60 mm
Minimum tubing ID above pump	70 mm
Length of rotor helix	1,2 m
Length of stator elastomer	1,0 m
Length between elastomer and rotor stop	16 mm
Stator thread specification	2 3/8 in
Rotor thread specification	1 1/16 in, API 3/4 in rod
Maximum speed of PCP	500 r/min
Head rating	12 MPa

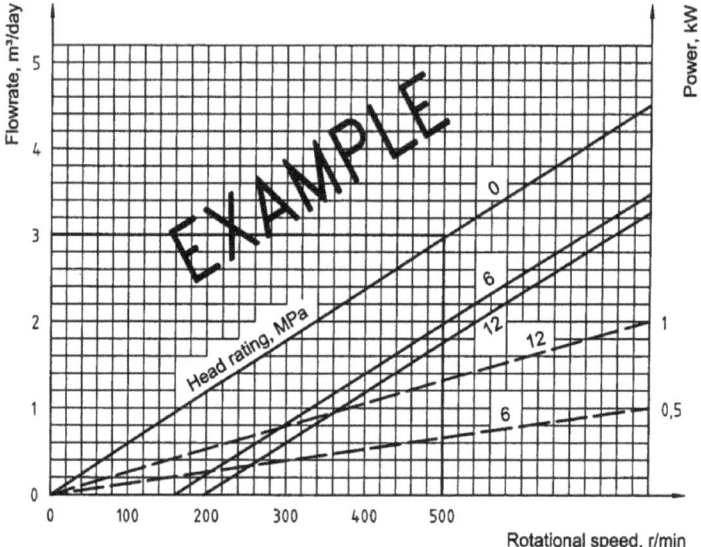

ISO 15136-1:2001(E)

Annex B
(normative)

PCP test report data sheet

Test characteristics		Order reference	
Pump intake pressure:	kPa	Purchase order No.:	
Fluid test temperature:	°C	Customer:	
		Rotor code:	
Date of test:		Stator, serial No.:	
Location of test:		Rotor, serial No.:	
		Elastomer name:	
Tested by:		Rotor coating:	

	N = r/min			q_C = m³/d				
Measured points	q_V m³/d	η_V %	Δp kPa	P_h kW	Γ N·m	P_{req} kW	η %	T °C
P1								
P2								
P3								
P4								

q_C actual pump displacement (m³/d), at zero discharge pressure, for a considered rotational speed, N (r/min);

q_V measured flowrate (m³/d), at a specific rotational speed (r/min) and differential pressure (kPa);

η_V volumetric efficiency, %: $(q_m/q_c) \times 100$;

Δp differential pump pressure, kPa: discharge pressure minus intake pressure;

P_h hydraulic pump power, kW;

Γ pump torque, N·m;

P_{req} power required to drive the pump, kW;

η pump overall efficiency, %: $(P_h/P_{req}) \times 100$;

T actual fluid test temperature, °C.

If the temperature of test is different from field temperature, the manufacturer shall explain and adjust the results; the same applies if the elastomer will swell when immersed in the oil.

NOTE 1   At a speed, N, at different points P1, P2, ... the values of pressure are measured; then the same operation can be repeated for another speed.

NOTE 2   Refer to ISO 31 for units.

ISO 15136-1:2001(E)

Annex C
(informative)

Application design specification data sheet

DATE: _____ COMPANY NAME: _____ CONTACT: _____
TELEPHONE No.: _____ FAX/e-MAIL No.: _____ WELL NAME: _____
FIELD: _____ WELL TYPE VERTICAL: _____ DIRECTIONAL: _____ SLANT: _____

PERFORATION DEPTH: TVD _____ m, measured _____ m
PUMP SETTING DEPTH: TVD _____ m, measured _____ m
TOTAL TUBING LENGTH: TVD _____ m, measured _____ m
FLUID LEVEL FROM SURFACE (TVD): STATIC: _____ m DYNAMIC: _____ m
PRODUCTIVITY INDEX: _____ m^3/d/kPa BOTTOMHOLE PRESSURE: _____ kPa
FLOWLINE PRESSURE: _____ kPa CASING PRESSURE: _____ kPa
CASING ID: _____ mm TUBING ID: _____ mm
TUBING THREAD TYPE AND SIZE: _____ WELLHEAD CONNECTION TYPE AND SIZE: _____
DRIVE STRING: TYPE: ____ OD: _____ mm COUPLING OD: ____ mm THREAD: ____

DAILY FLUID PRODUCTION: _____ m^3/d   PROJECTED DAILY FLUID PRODUCTION: _____ m^3/d
CURRENT MODE OF PRODUCTION: _____
WATER CUT: _____ %
SOLIDS CONTENT: _____ % _____
GAS/OIL RATIO: __ m^3/m^3 GAS BUBBLE POINT: _____ kPa GAS SPECIFIC GRAVITY: _____ S.G.
VISCOSITY: _____ cP at _____ °C
API OIL GRAVITY: _____ MOLE PERCENT LIGHT MONOAROMATIC: _____ %
TOTAL FLUID SPECIFIC GRAVITY: _____ OIL: ____ WATER: _____
H_2S: _____ %   pH: _____
CO_2: _____ %   CHLORIDES: ____ %
PUMP INLET TEMPERATURE: _____ °C WELLHEAD TEMPERATURE: _____ °C
AMBIENT TEMPERATURE: _____ °C ELECTRICAL ZONE CLASSIFICATION: _____

PRIME MOVER TYPE: _____ GAS _____ ELECTRIC
VOLTAGE _____ V   FREQUENCY: _____ Hz
POWER TRANSMISSION MODE: _____ DIRECT DRIVE _____ HYDRAULIC

CHEMICAL TREATMENT PROGRAMME: YES _____ NO _____ IF YES THEN DESCRIBE: _____
_____

COMMENTS: _____
_____

Bibliography

Arnaudeau M (1978) *"Les pompes Moineau"*. Forages 80.

Arrellano J (1996) *"In situ PC Pump Testing* 1996 Progressing Cavity Pump". Workshop. November 19, Tulsa, OK.

Bagin LA and Gorbatov VS (1987) *"Problems and Prospects of the Implementation of Electrical Downhole Screw Pumps for the Production of Very Viscous Petroleum"*. Nefjanoe Khozjajstvo, 8.

Baldenko DF and Baldenko FD (1996) *"Multilobe Progressing Cavity Pump. Operating Process Particularities, Geometrical Parameters Optimization and Field of Application"*. 1996 Progressing Cavity Pump Workshop, November 19. Tulsa.

Baldenko DF, Baldenko FD and Shmidt AP (1997) *"Screw Downhole Motors: News Designs and Control Methods"*. VNIIBT.

Bauquin J-L, Boireau C, Lemay L and Seince L (2005) *"Development Status of a Metal Progressing Cavity Pump for Heavy-Oil and Hot-Production Wells"*. SPE 97796 (presented at the 2005 SPE/PS-CIM/CHOA International Thermal Operations and Heavy Oil Calgary Alberta 1-3 November 2005.

Beauquin J-L, Chalier G & Ndinemenu F TOTAL SA, Lemay L & Seince L PCM and Jahn S KUDU (2008) *"Metal PCP Field Trial Pushes up Pumping Window for Heavy Oil Hot Production: Joslyn Field Case, WHOC 2008-498*, World Heavy Oil Congress Edmonton Alberta Canada 10-12 march 2008.

Bourke JD (1996) *"Compensating Eccentric Motion in Progressing Cavity Pumps"*. World Pumps. May 1996.

Bratu C, Multiphase PCP's rotor and stator principle, patent FR 2,865,781 filed Jan 30, 2004, Bratu C Assigee: PCM *"Progressing Caviy Pump"* US 7,413,416B2 Aug 19, 2008.

Bratu C (2005) *"Progressing Cavity Pump Behaviour in Multiphase Conditions"*. SPE 95272. Annual Technical Conference and Exhibition, Dallas, USA, 9-12 October 2005.

Bratu C SPE and L Seince PCM (2005) *"Progressing Cavity Pump for Multiphase and Viscous Liquid Production"*. SPE 97833 International Thermal Operations and Heavy Oil Symposium Calgary, Alberta, Canada, 1-3 November 2005.

Campbell B, MacKinnon J, Bandy TR and Hampton T (1996) *"Optimizing Artificial Lift Operations Through the Use of Wireless Conveyed real Time Bottom Hole Data"*. Paper SPE 36596. 6-9 October 1996.

Centrilift-Hugues *"Submersible Pump Handbook – The System – Step VIII – Gas Separator Calculation"*.

Cholet H (1982) *"Essais en puits expérimental de pompes de fond prototypes"*. ARTEP/Moineau. IFP, Ref. 30 798, December 1982.

Cholet H, Trocquemé FJ et Garraux J (1985) *"Dispositif utilisable notamment pour le pompage d'un fluide très visqueux et/ou contenant une proportion notable de gaz particulièrement pour la production du pétrole"*. FR Patent No. 2 551 804. March 15.1985. Patent US No. 4 718 824, January 12, 1988. + GB.

Cholet H et Chanton E (1986) *"La pompe Rodemip"*. Forages No. 111.

Cholet H et Chanton E (1989) *"Dispositif de pompage d'un fluide dans le fond d'un puits"*. FR Patent No. :2 617 534, January 6, 1989. Patent US No. 4 957 161, September 18, 1990. + BE, EP, GB, IT. DE.

Cholet H, Lessi J. (1989) *"Dispositif de pompage d'un fluide au fond d'un puits notamment à zone basse fortement inclinée ou horizontale"*. Fr patent No 2631 379, Novembre 17, 1989.

Cholet H (1991) *"Dispositif de pompage d'un liquide et en particulier d'un liquide à forte viscosité"*. FR Patent ~o. 2 656 035. June 21, 1991.

Cholet H (1995) *"Dispositif de séparation d'un mélange de gaz et de liquide à l'admission d'une pompe au fond d'un puits fore"*. FR Patent No. 2 656 652. August 25,1995. Patent US 5 113 937; May 19. *1992.*+ DK, EP, GB, IT.

Cholet H et Lessi J (1993) *"Dispositif et méthode de pompage d'un liquide visqueux comportant l'injection d'un produit fluidifiant. Application aux outils horizontaux"*. FR Patent No. 2 692 320, December 17. 1993. Patent US No 5 348 094. Sept 20, 1994.

Cholet H et Bourdon JC (1994) *"Système de pompage comportant une pompe volumétrique à grand debit"*. FR Patent No 2 696 792, April 15. 1994, FR Patent No 2 108 315, April 14, 1994.

Cholet H et Petit H (1995) *"Système de pompage comportant une pompe volumétrique à grand débit"*. FR Patent No. :2, 714120. June 23, 1995. CA Patent No 2 138 356. June 17, 1995.

Cholet H (1996) *"Méthode et système de pompage comportant une pompe volumétrique entrainée par un tube continu. Application aux puits déviés"*. FR Patent No 2 727 475. May 31, 1996.

Cholet H et Wittrisch C (1997) *"Méthode et dispositif de production par pompage dans un drain horizontal"*. FR Patent No. 2741 382, May 23, 1997.

Cholet H, Vandenbroucke E, PCP metal stator *"Progressive cavity pump with composite stator and manufacturing process"* FR 2,794,498 filed 07 June.7, 1999, US 6,336,796 Jan.8, 2002.

Cholet H, Wittrisch C. Bootsmann (1996) Gas anchor, *"Méthode et dispositif de production par pompage dans un drain horizontal"* FR Patent No. 2741 382, May 23, 1997, "Method ans device for producing by pumping in horizonatl well" US 5,829,529 Nov 3, 1998.

Clegg JD, Bucaram SM and Hein N.W.Jr. (1993) *"Recommendations and Comparisons for Selecting Artificial Lift Methods"*. Journal of Petroleum Technology, December.

Deltassand M (1996) *"Designing PC Pumps for Abrasive Applications"*. 1996 Progressing Cavity Pump Workshop, November 19, 1996, Tulsa. OK.

Dunn L, Mathew C and Zahacy TA (1995) *"Progressing Cavity Pumping System Applications in Heavy Oil Production"*. Paper SPE 30271, June 19-21. 1995.

Elf-Aquitaine (1984) *"Essai d'une pompe de fond Aremip sur le puits de Laméac"*. Mission France 1983. *Forages 105.*

EMIP (1993) *"Application des pompes Rodemip au développement d'un champ"*. *Forages 140.*

Guerra E and Sanchez A from Sherritt International Oil and Gas and Matthews C from C-Fer Technologies (2009) *"Field Implementation Experience With Metal PCP Technology in Cuban Heavy-Oil"*. SPE 120645 prepared for 2009 SPE Production and Operations Symposium, Oklahoma City april 2009.

ISO/TC 67/SC 4/WG 4 Working group (1997). ISO CD 15136: *"Standard for Progressing Cavity Pump Systems for Artificial Lift in the petroleum industry"*.

Haworth C (1996) *"Updated field case studies 071 application and performance of bottom drive progressing cavity pumps"*. 1996 Progressing Cavity Pump Workshop. November 19, 1996, Tulsa. OK.

Huc A-Y, (2011) *"Heavy Crude Oils from Geology to Upgrading an Overview"*. IFP Energies nouvelles Publications, Editions TECHNIP.

Lea JF, Anderson PD and Anderso DG (1987) *"Optimization of Progressive Cavity Pump Systems in the Development of Clearwater Heavy Oil Reservoir"*. Paper 87-38-03 presented at the 1987 Petroleum society CIM, Calgary, June 7-12.

Lesage J (1993) *"Les élastomères dans les industries pétrolières et parapétrolière"*. revue IFP 48, 4.

Loktev AV, Bolgov ID, Semin VG et Michurin VG (1995) *"L'utilisation des pompes à vis à commande en surface dans la SA"*. Chernogorneft, Neft Khoz. 9".

Olivet A, Gamboa J and Kenyery F (2002) *"Understanding the Performance of a Progressive Cavity Pump with Metallic Stator"*. SPE 77730 (presented at the 2002 SPE ATCE conference, San Antonio, 29 Sept. – 2 Oct.).

Matthew CM and Dunn LJ (1996) *"Drilling and Production Practices to Mitigate Sucker Rod/Tubing-wear-Related Failures in Directional Wells"*. Paper SPE 22852, production & facilities; November 1993.

Matthew CM (1996) *"Instrumentation PC Pump Charges Pump"*. 1996 Progressing Cavity Pump Workshop, November 19, 1996, Tulsa, OK.

McCoy JN, Podio AL and Woods MGuillory C (1996) *"Field Test of a Decentralized Downhole Gas Separator"*. paper SPE 36599, 6-9 October 1996.

McCoy JN (1996) *"Analysis and Optimization of Progressing Cavity Pump Systems by Total Well Management"*. 1996 Progressing Cavity Pump Workshop, November 19, 1996 Tulsa. OK.

Mills RA (1994) *"Progressing Cavity Oilwell Pumps – Past-present and Future"*. Journal of Canadian Petroleum Technology 33,4.

Moineau RJ (1932) *"Pompe"*. FR Patent No 695 539 September 30, 1930. US Patent US No 1 892 217, December 27, 1932.

Moineau RJ (1935) *"Mécanisme à engrenages perfectionnés, utilisables comme pompe, compresseur, moteur ou simple dispositif de transmission"*. FR Patent No 787 711, July 8, 1935.

Moineau RJ (1937) *"Gear Mechanism"*. Patent US No. 2 085 115, June 29, 1937, Reissued February 27, 1940 – Re.21 374.

Moineau R.J (1949) *"Helical Multiple Pump"*. Patent US No 2 483 370, September 27, 1949.

Nelik Lev, PE, APICS and Jim Brennam *"Progressing Cavity Pumps, Downhole, Pumps, and Mudmotors"*. Gulf publishing Company, Houston 2005.

Nguyen JP and Gabolde G, IFPEn, *"Drilling Data Handbook"*. Editions Technip 1999 and

Niels Aage, and working group, *"Mathematical Problems for Moineau Pumps"*. November 6, 2006 moineau pumps grundfos.

Parise N, Lehman M and Amara A-B (2009), PCM *"Vulcain™ Metal-to-metal PCP Artificial Lift Systems"* (presented at the 2009 MEALF conference, Barain).

Perrin D, Caron M et Caillot G (1995) *"La production fond"*. Edition Technip.

Ratov AM and Khejfec JS (1979) *"General Information, Pump Manufacturing Range KHM-4. The downhole Mono-helical Electrical Pumps in USSR and Abroad"*. Moskva-Cintikhimneftemash.

Rae, M Shell Canada, Seince L PCM, Mitskopoulos M and Kudu (2011) *"All Metal Progressing Cavity Deployed in SAGD"*. World Heavy Oil Congress Edmonton Alberta Canada 2011.

Ramos MA, Brown JC, Quijada M, Parra R, Romero J, Petróleos de Venezuela, SA and Seince L, PCM (2008) *"Experiences and Best Practices in the Use of PCP's in Orinoco Belt Carabobo Area, Venezuela"*. Heavy Oil Congress Edmonton 10-12 march 2008.

Revard JM (1995) *"The Progessing Cavity Pump Handbook"*. PennWell Oub Co.

Rondy P, Cholet H and Federer I (1993) *"Optimization of Heavy Oil and Gas Pumping in Horizontal Wells"*. Paper SPE 26555, 3-6 October 1993.

Saveth KJ and Klein ST (1989) *"The Progressing Cavity Pump: Principle and Capabilities"*. Paper SPE 18873, Production Operations Symposium, March 13-14,1989.

Seince L, PCM Canada, D,Caballero, PCM, and N, Chacin, Equimavenca, *"Multiphase Progressive Cavity Pumps Operated in Harsh Conditions"*. SPE 137 168 12-14 September 2010 Alberta Canada.

Tiraspolsky W (1985) *"Hydraulic Downhole Drilling Motors"*, Editions Technip, Paris.

Tupman J (1996) *"Progressing Cavity Pumps in Open Hole"*. 1996 Progressing Cavity Pump Workshop, November 19, 1996, Tulsa, OK.

Waggs BT *"Progressing Cavity Pump Inspection and Reporting"*: 2007-C-FER Technologies.

Wittrisch C *"Injection through PCP rotor to the horizontal drain hole"* *"Méthode et système de pompage dans un puits pétrolier"* FR 2 859 753 filed 16/09/03, *"Method and system for pumping an oil well"*, US 7.290.608 Nov, 6, 2007.

Wright DW and Adair RL (1993) *"Progressive Deliver Highest Mechanical Efficiency/Lowest Operating Cost in Mature Permian Basin Water Flood"*. Paper SPE 25417, March 21-23.

Zabel L (1996) *"Electrical Submersible Progressing Cavity Pump ESPC, an Alternative Lift Method for Problem Applications"*. 1996 Progressing cavity Pump Workshop, November 19, 1996, Tulsa.

Mathematical problems for Moineau pumps

Yves van Gennip, Department of Mathematics UCLA

Working group: Niels Aage, John Donaldson, Yuyang Feng, Yves van Gennip, Helge Grann, Jens Gravesen, Andriy Hlod, Troels Steenstrup Jensen, Anders Astrup Larsen, Kamyar Malakpoor, Steen Markvorsen, David Moreno, Jos in't panhuis, Peter in't panhuis, Peter Røgen, and Erwin Vondenhoff. November 6, 2006.

The geometry of the Moineau pump, a mathematical analysis of the pump

Jens Gravesen, Technical University of Denmark, Department of Mathematics 26 March 2008. Work with the Danish pump manufacturer Grundfos and Technical University of Denmark, presented at the 57th European Study Group with Industry in Denmark.

List of figures

Figure 1	Moineau Deauville competition (source: *Musée de l'Air et de l'Espace*)	page 3
Figure 2	René Moineau's "Capsulism" pump (source: *PCM*)	page 3
Figure 3	René Moineau's life 1887-1948 (source: *Edition de l'Officine*)	page 5
Figure 4	Single lobe PCP type 1-2, rotor and stator view (source: *National Oil Well*)	page 10
Figure 5-1	Operating principle diagram of PCP 1-2 (source: *PCM*)	page 11
Figure 5-2	PCP 1-2 operating principle (source: *Petrobras*)	page 12
Figure 6-1	Pump type 3-4: Hypocycloids H_1 and H_2 (source: *PCM*)	page 13
Figure 6-2	Pump type 3-4 Hypocycloid envelopes (source: *PCM*)	page 13
Figure 6-3	Hypocycloid profile cross sections	page 14
Figure 7	Pump 1-2 geometry (source: *PCM and IFP Energies nouvelles*)	page 16
Figure 8-1	Rotor-Stator perspective	page 17
Figure 8-2	Mono lobe (1-2) Rotor and stator geometry (source: *IFP Energies nouvelles*)	page 17
Figure 9	Multi lobe pump relationship (source: *M.T. Gusman and D.F. Baldenko*)	page 20
Figure 10	Multi lobe pumps geometry example (source: *Netzsch*)	page 21
Figure 11	PCP typical configuration (source: *Protex*)	page 23
Figure 12	PCP efficiency versus rotor fit and pump life (source: *OilLift Technology Inc*)	page 36
Figure 13-1	Definition Heavy Oil, Extra-Heavy Oil (source: *API & Canadian Center of Energy*)	page 38
Figure 13-2	Specific gravity/API Conversion	page 38
Figure 13-3	Composition of Extra-Heavy Oil and Light Oil (source: *IFP Energies nouvelles*)	page 39
Figure 14-1	Temperature effect on oil viscosity (source: *From Owens and Souler*)	page 40
Figure 14-2	Heavy oil/kerosene dilution and viscosity reduction (source: *IFP Energies nouvelles*)	page 41
Figure 15	Flow rate and PCP position in an oil well containing gas (source: *IFP Energies nouvelles*)	page 49
Figure 16	Recommended limitations of working for wells containing sand (source: *PCM*)	page 51
Figure 17	Relationship between the rotating speed and the fluid level (source: *IFP Energies nouvelles*)	page 53
Figure 18	PCP performance curves (source: *ISO15136-1*)	page 58
Figure 19-1	Example of pump manufacturer specification sheet (source: *OilLift Technology Inc*)	page 64

Figure 19-2	Example of pump manufacturer specification sheet (source: *PCM*)	page 65
Figure 19-3	PCP data sheet well completion and production (source: *IFP Energies nouvelles*)	page 67
Figure 20-1	Flow pattern in vertical wells (source: *A.E. Dukler. University of Houston*)	page 72
Figure 20-2	Flow pattern in horizontal pipes (source: *A.E. Dukler. University of Houston*)	page 72
Figure 21	Static gas separator (source: *IFP Energies nouvelles/Horwell*)	page 74
Figure 22	Centrifugal gas separators (source: *CANAM, Evolution Oil tools*)	page 75
Figure 23	Gas separator systems for horizontal wells (source: *IFP Energies nouvelles*)	page 76
Figure 24	Non-dimensional gas liberation curve (source: *Nabla Corporation*)	page 78
Figure 25	Relation between fluid level and PCP performance (source: *C.S. Resources/IFP Energies nouvelles*)	page 80
Figure 26	Solid drive rods (source: *Tennaris*)	page 81
Figure 27	Drive rod and NON rotating centralizers (source: *PCM and Canam*)	page 87
Figure 28	Drive string/Tubing contact geometry (source: *IFP Energies nouvelles*)	page 89
Figure 29	Centralizer location in relation to side force versus well profile (source: *PCM*)	page 90
Figure 30	Hollow PCPRods (source: *Tennaris*)	page 92
Figure 31	Continuous drive rod and transport (source: *Weatherford*)	page 94
Figure 32-1	Tubing anchor catcher fixed to the PCP stator (source: *Halliburton*)	page 96
Figure 32-2	Torque anchor under PCP stator (source: *Evolution Oils Tools Inc*)	page 97
Figure 33	Drive head with solid shaft (source: *PCM*)	page 102
Figure 34	Drive head with hollow shaft and return angle drive (source: *Griffin Pumps*)	page 103
Figure 35-1	Rod twist before pump stars (source: *Zenith Oilfield tech*)	page 104
Figure 35-2	Drive head with back-spin brake rotor (source: *KUDU*)	page 105
Figure 36	Examples of drive systems (source: *PCM*)	page 107
Figure 37-1	Belt drive head (source: *Oil Lift Technology*)	page 111
Figure 37-2	Drive head for slant well Source (source: *Kudu*)	page 112
Figure 37-3	Electric and hydraulic drive head (source: *GrenCo*)	page 112
Figure 37-4	Hydraulic drive head (source: *KUDU*)	page 113
Figure 37-5	Hydraulic vertical drive head (source: *Weatherford*)	page 113
Figure 37-6	Moyno hydraulic Gear Drivehead (source: *R&M Energy Systems*)	page 114
Figure 37-7	Gear drive (source: *OilLift Technology Inc*)	page 114
Figure 38-1	Direct Drive PCP head with permanent magnet motor (source: *Daqing Jinghong Petroleum Equipment Factory*)	page 115
Figure 38-2	Direct Drive PCP head with permanent magnet motor (source: *Source Shengli Electric limited Company*)	page 115
Figure 38-3	Direct Drive PCP head with permanent magnet motor (source: *Torquedrive head from Advantages*	page 116
Figure 39	Rotating seal Stuffing Box	
	DuraSeal-500 Rotating Stuffing Box (source: *Weatherford*)	page 118
	Zero leak stuffing (source: *CAN-K*)	page 118

List of figures

Figure 40	Rod Lock BOP (source: *OilLift Technology Inc*)	page 119
Figure 41	RODEC Tubing Rotator Spool (source: *RODEC/R&M*)	page 119
Figure 42	Running in the rotor (source: *IFP Energies nouvelles*)	page 123
Figure 43	Running in rods (source: *IFP Energies nouvelles*)	page 124
Figure 44-1	Basic principle Insert progressing cavity pump (source: *IFP Energies nouvelles*)	page 130
Figure 44-2	Insert PCP Systems (source: *KUDU*)	page 131
Figure 45	Electrical Submersible PCP Schematic principle (source: *Baker Centrilift*)	page 134
Figure 46-1	RedaMAX ES flat or round power cable (source: *Schlumberger*)	page 137
Figure 46-2	Examples of cable clamp protectors	
	Composite material cable protector (source: *BossCLAMP Artificial Lift company*)	page 138
	Cast steel cross coupling cable (source: *S.U.N. Engineering*)	page 138
Figure 47	Well head with power feed thru connector (source: *ITTCannon/BIW*)	page 138
Figure 48	Permanent Magnet Electric Motor Principle	page 140
Figure 49	Multiphase PCP rotor and stator principle: patent (source: *Bratu*)	page 149
Figure 50-1	PCP pump metal stator and rotor (source: *PCM*)	page 152
Figure 50-2	PCM Vulcain MET Pump characteristics (source: *PCM*)	page 153
Figure 51	Solid metal stator (source: *Robbins&Myers Energy Systems*)	page 154
Figure 52	PCP metal stator principle: patent (source: *IFP Energies nouvelles*)	page 154
Figure 53	Uniform thickness stator elastomer and standard stator (source: *EUROMax PC pumps*)	page 155
Figure 54	Hollow rotor (source: *Kachelle*)	page 156
Figure 55	Injection through rotor to the horizontal drain hole: patent (source: *IFP Energies nouvelles*)	page 157
Figure 56-1	Drive head with a lateral injection swivel port (source: *KUDU*)	page 157
Figure 56-2	Hollow drive head injection and coiled tubing drive string (source: *IFP Energies nouvelles*)	page 158
Figure 57	Hydraulic lines to hydraulic drive PCP (source: *CJS*)	page 161
Figure 58	Steam injection through the PCP hollow rotor principle (source: *Kachelle*)	page 164
Figure 59	Cold production with diluents through injection line (source: *IFP Energies nouvelles*)	page 165
Figure 60	Downhole sensor on tubing (source: *GRC*)	page 171
Figure 61	Surface pressure safety switch	page 172
Figure 62-1	SAM™ PCP Well Manager Automation and algorithm (source: *KUDU Industries and Lufkin Automation*)	page 173
Figure 62-2	Vector Flux Drive and monitoring read out (source: *Weatherford*)	page 173
Figure 63-1	The crest of the rotor wear (source: *C-Fer*)	page 182
Figure 63-2	High GOR producer elastomer bubbles and rupture (source: *Baker*)	page 183
Figure 63-3	Elastomer damaged by high discharge pressure and heat (source: *Baker*)	page 183
	Web Sites and manufacturers (source: *IFP Energies nouvelles*)	page 199-201

Index

A
Abrasion 68
API degrees 37

B
Back spin energy 102
Bubble point level 52

C
Cable clamp protectors 137
Calculated flow rate 26
Cavities 18, 19
Coiled tubing drive 93
Continuous drive rod 94
Corrosion 68

D
Data communications 169
Data sheet 63, 70
Diameter (D) 18
Diameter (E) 16
Dilution 40
Direct drive head 109, 115, 116
Downhole hydraulic motor 161
Drive head 22, 101, 106, 109, 111, 116
Drive head brake 105
Drive rods 81
Dynamic level 52

E
Eccentricity (E) 12, 15, 18
Efficiency 33, 34
Elastomer stator 29, 31

Electrical submersible PCP 133
ES power cable 136
ES-PCP 134
ES-PCP advantages 143
ES-PCP retrivable 140
Extra-heavy oil 38

F
Fit 35
Flow line back head pressure 55
Flow rate 26, 44, 57
Fluid data 68
Fluid temperature 50

G
Gas at the pump inlet 71
Gas content 48
Gas liquid ratio 48
Gas oil ratio 48, 77
Gas separator 73
Gravity 37
Guideline and identification 59

H
Head loss 55
Head of pressure 56
Head rating 26, 52, 57
Head, true vertical depth 55
Heavy crude 37
Hollow rods 84, 91
Hollow rotor 156
Hydraulic lines 161
Hydraulic power 106
Hypocycloid 14

I

Injection of diluents 165
Insert PCP 129
ISO 15136-1 29, 60, 203

K

Kinematics 16, 25

L

Leak rate 26
Light crude 37
Load 28

M

Major helix rotor 18
Medium crude 37
Model identification (PCP) 61
Monitoring 167, 169
Mono-lobe pump 15, 194
Multi-lobe pump 19, 193

N

Natural bitumen 38
Net lift 55
Net positive suction head 55
Non rotating centralizer 87

O

Operating problems 184

P

PCP advantages 187
PCP failures 181
PCP high capacities 147
PCP high temperature elastomer 151
PCP limitations 188
PCP metal stator 151
PCP metal stator Vulcain 152
PCP multiphases 148
PCP software 177
Permanent magnet motor 109, 115, 116, 139
Pitch 16, 18

Polished rod 95
Pressure head 55
Production fluid data 68
Productivity index 44
Pump intake 55

R

Resistant torque 85
Rod lock bop 118
Rod torque 84
Rotary seal 117
Rotating speed 28, 58
Rotor 10, 21
Rotor failures 181
Rotor lobes number 19

S

Safety valve 99
Sand transportation 50
Slippage 33, 34
Solid drive rod 81
Solid metal stator 154
Solid rods 84
Specific gravity 37
Specific gravity/API Conversion 38
Stator 10, 22
Stator elastomer failures 182
Stator lobes number 19
Steam injection 163
Stuffing box 117, 185
Submergence level 52

T

Tag bar screen 98
Tag bar-stop pin 98
Temperature 66, 192
Thrust 85
Torque 27, 29
Total net lift 55, 195
True vertical depth 55
Tubing anchor 95
Tubing anchor catcher 95
Tubing rotator 119
Tubing torque anchor 96

U

Uniform thickness stator 155

V

Variable speed drive head 108, 160
Viscosity 39
Viscosity and temperature 39

Viscosity reducer 48
Volume displacement 26

W

Well management software 172
Well monitoring 189
Wellhead pressure 52, 54